Oxford Chemistry Series

General Editors

P. W. ATKINS   J. S. E. HOLKER   A. K. HOLLIDAY

Oxford Chemistry Series

J. MANN

Professor of Organic Chemistry, University of Reading

Secondary metabolism

Second edition

Clarendon Press · Oxford ·

Oxford University Press, Walton Street, Oxford OX2 6DP
Oxford New York Toronto
Delhi Bombay Calcutta Madras Karachi
Petaling Jaya Singapore Hong Kong Tokyo
Nairobi Dar es Salaam Cape Town
Melbourne Auckland
and associated companies in
Berlin Ibadan

Oxford is a trade mark of Oxford University Press

Published in the United States
by Oxford University Press, New York

First edition 1978
Reprinted with corrections in paperback 1980
Second edition published 1987
Reprinted (with corrections) 1992

British Library Cataloguing in Publication Data
Mann, J.
  Secondary metabolism.
  (Oxford chemistry series: 33)
  Bibliography: p.
  Includes index.
  1. Metabolism, Secondary. I. Title. II. Series
QH521.M36 1986 574.19'24 86–5390
ISBN 0–19–855530–X
ISBN 0–19–855529–6 (pbk)

Library of Congress Cataloging in Publication Data
Mann, J.
  Secondary metabolism.—2nd ed.—
  (Oxford Chemistry Series, 33)
  1. Metabolism, Secondary 2. Natural
products
  I. Title
  574.19'24
  ISBN 0–19–855530–X
  ISBN 0–19–855529–6 Pbk

Printed in Great Britain by St Edmundsbury Press,
Bury St Edmunds, Suffolk

Editor's foreword

The chemistry of natural products has always excited the interest of the scientist and it provided the early stimulus for the foundation of organic chemistry as a separate discipline. The rapidly growing knowledge of natural-product structure led naturally to questions concerning the transformations and functions of biological molecules and hence to the subject of biochemistry.

Traditionally biochemists have studied the essential life processes of primary metabolism while the organic chemist has tended to concentrate his interests on secondary metabolites, where the problems of structure elucidation provided the necessary initial intellectual challenge. Great progress has been made in both areas and a composite picture of biosynthesis has emerged of sufficient clarity to satisfy the differing philosophies of biochemist and organic chemist alike. The subject now provides an essential bridge between the physical and life sciences.

The remarkable advances which have been made in the understanding of biosynthesis have depended on specialized techniques drawn from a wide range of disciplines. The use of mutants, isolated enzyme systems, radioactive tracer techniques, and, more recently, the application of n.m.r. methods to metabolites enriched with precursors containing stable isotopes, have all contributed to the picture. The subject is now so important that biologist, biochemists, and organic chemists must be exposed to this exciting field of study.

Dr Mann is an organic chemist with a very strong interest in natural-product chemistry and is well equipped to present a balanced picture of the field. Although the book is written primarily for the organic chemist, and provides the basic knowledge of biosynthesis essential to the armoury of any organic chemist, the material covered will also be of great interest to the biochemist and biologist. Indeed, any scientist who is concerned with the chemistry of life processes will discover in this book an account of the remarkable progress that has been made in the last twenty years.

J. S. E. H.

For Sara, Sebastian,
Cressida, and Octavia

Preface to the second edition

During the eight years since the first edition was published, sophisticated n.m.r. techniques (especially observation of isotope-induced shifts when ^{13}C and ^2H or ^{18}O isotopes are jointly present in the administered substrates), and the use of isolated enzymes, have revolutionized the study of biosynthetic pathways. In particular, subtle details of these routes may now be identified, and the biogenesis of ever more complex natural products can be probed. These innovations are highlighted in this completely updated edition, though the emphasis is still on a consideration of the main pathways of biosynthesis, and on the biological activities and ecological significance of secondary metabolites. In order to include as much new material as possible, without complicating the main part of the text, some of the more exotic or complex compounds have been included in the problems section (answers are now provided), or are referred to in the list of research references.

Reading   J. M.
January 1986

Preface to the first edition

The chemistry of natural products has been studied seriously for about one hundred and fifty years; but investigation of their biosynthesis is a much more recent research endeavour, and has been greatly facilitated by the advent of modern spectroscopic techniques and the availability of isotopically labelled compounds. The results of these investigations—the proven pathways of biosynthesis and interconversion of natural products—are the subject of this book.

In the first chapter the idea of primary and secondary metabolic pathways is introduced. These produce, respectively, essential and ubiquitous compounds (e.g. carbohydrates, proteins, and nucleic acids), and apparently non-essential compounds, or secondary metabolites. This latter group is then classified according to the small precursor molecules (or 'building blocks') from which they are derived, and most of the remainder of the book is concerned with a detailed discussion of the various biosynthetic pathways. The final chapter considers ecological chemistry, and attempts to show how secondary metabolites are used as mediators of complex interactions between different species.

My aim was to write a text that was suitable for undergraduate courses in natural products/secondary metabolism, and of use to research workers entering the field, who required a general introduction to the subject. It is not a reference text, and does not cover the chemistry of secondary metabolites in any great detail: there are many excellent texts which serve these purposes. However, all major classes of secondary metabolites are dealt with, and their biosynthesis is covered in some detail. In the main, only well-established results are included, but at the end of each chapter there are questions, whose answers may be found in the recent research literature.

Finally, I have stressed the biological properties of these compounds and their pharmacolgy, toxicology, and ecological significance, because I feel these aspects have received scant attentions in the past. Secondary metabolites can no longer be thought of as the useless detritus produced by obscure metabolic pathways: many have a profound influence on the interactions of one species with another. When we understand these interactions a little better, we should be able to ensure that we do not disturb the environment to our own detriment or to that of the species with which we coexist.

I should like to thank Dr Holker for his invaluable help and guidance, and also Dr Derek Banthorpe, Dr Jeffrey Harborne, and Dr David Crout, who read parts of the manuscript. Their comments and criticism were of great assistance.

Reading   J. M.
1977

Contents

Contents

Abbreviations used and stereochemical nomenclature

Abbreviations

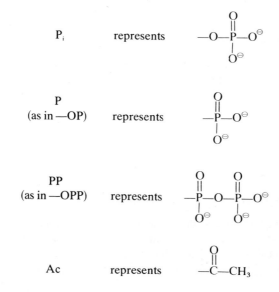

P_i	represents	$-O-\overset{\overset{O}{\|\|}}{\underset{\underset{O^{\ominus}}{\|}}{P}}-O^{\ominus}$
P (as in —OP)	represents	$-\overset{\overset{O}{\|\|}}{\underset{\underset{O^{\ominus}}{\|}}{P}}-O^{\ominus}$
PP (as in —OPP)	represents	$-\overset{\overset{O}{\|\|}}{\underset{\underset{O^{\ominus}}{\|}}{P}}-O-\overset{\overset{O}{\|\|}}{\underset{\underset{O^{\ominus}}{\|}}{P}}-O^{\ominus}$
Ac	represents	$-\overset{\overset{O}{\|\|}}{C}-CH_3$

Stereochemical nomenclature

It is assumed that the reader is familiar with the Cahn–Ingold–Prelog (R,S) nomenclature, but two other types of stereochemical nomenclature that are commonly encountered in discussions of secondary metabolism are given below.

Prostereoisomerism

When deuterium or tritium isotopes are introduced into organic molecules, new chiral centres are often produced. So, for example, in the reduction of ethanal by sodium borodeuteride, two stereoisomers of ethanol are produced:

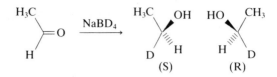

Upon oxidation either two hydrogens or a hydrogen and a deuterium may, in principle, be lost, to yield monodeutero ethanal or ethanol. In ethanol itself, the two methylene hydrogens are apparently identical, though an oxidation mediated by an enzyme will usually be stereospecific, and only one of the two possible hydrogens will be lost. In order to distinguish between these atoms we can call them pro-R and pro-S (H_R and H_S), implying that if the former atom is replaced by a deuterium or tritium atom, the molecule will then have R-stereochemistry, and if the latter is replaced the S-stereochemistry will result, viz:

Enantiotopic faces

Prochirality can be extended to the faces of molecules, and this is also of prime importance when discussing the stereochemistry of enzyme-mediated reactions. Thus, attack of a nucleophile at a carbonyl produces different chiral products, depending upon the direction of attack:

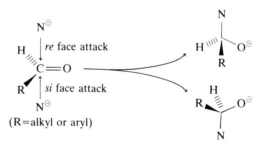

To distinguish the two faces, one considers the 'view seen' by the attacking species: if the groups are arranged (in the priority sense of the Cahn–Ingold–Prelog rules) in a clockwise direction this is then the *re* (rectus) face, and if in an anticlockwise direction this is then the *si* (sinister) face.

Clearly, the above aspects of stereochemistry apply equally well to asymmetric organic synthesis, as they do to secondary metabolism and biochemical processes in general.

1. Introduction

Natural products

This book is primarily concerned with the formation, structure, and biological activity of natural products. The term 'natural product' is commonly reserved for those organic compounds of natural origin that are unique to one organism, or common to a small number of closely related organisms. In most instances they appear to be non-essential to the plant, insect, or microorganism producing them, in marked contrast to the other organic compounds in Nature, sugars, amino acids, nucleotides, and the polymers derived from them, which are both essential and ubiquitous.

Two examples of typical natural products will help to clarify the definition. Morphine only occurs in two species of poppy, *Papaver somniferum* and *P. setigerum*, and although widely used and abused by Man, has no known function in these plants. Similarly, penicillins are produced by a few species of fungi, and by no other organisms. They have great value as antibiotics in the service of Man, but appear to serve no useful purpose in the microorganisms that produce them.

Man has used natural products, albeit as crude plant extracts, since the dawn of time and we still possess 'recipes' from mediaeval times. Thus Ambroise Paré (1517–90), a French surgeon of some note, treated gunshot wounds with a concoction of chamomile, melilot flowers, lavender, rosemary, sage, thyme, and the extract from red roses boiled in white wine. Thomas Sydenham (1624–89), a Bachelor of Medicine (and friend of Robert Boyle) prescribed finely powdered Peruvian bark (from the cinchona tree) mixed with syrup of cloves as a remedy for malaria. Quinine is a major component of the bark of the cinchona tree.

Primitive man found these extracts efficient as medicines for the relief of pain or alleviation of the symptoms of disease, as poisons for use in warfare and hunting, as effective agents for euthanasia and capital punishment, and as narcotics, hallucinogens, or stimulants to relieve the tedium, or alleviate the fatigue and hunger in his life. He must also surely have used the more odiferous and spicy compounds to obscure the odour of unwashed humanity, and to disguise the putrid or bland flavour of his food. Some of the compounds that he unwittingly employed are shown in Fig. 1.1.

Many of these natural products are still used today, and usually for the same general purpose. Curare, a plant extract which contains several toxic alkaloids, was used by South American Indians as an arrow poison since they discovered that it could paralyse even quite large animals. A

Medicines

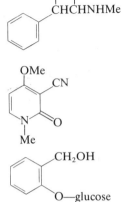

Ephedrine (respiratory ailments)

Ricinine (Castor oil: purgative)

Salicin (from willow bark: used by 'rural folk' for fevers; aspirin is a synthetic analogue)

Poisons

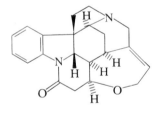

Strychnine

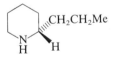

Coniine (hemlock)

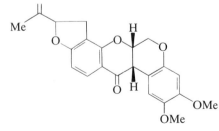

Rotenone (fish poison and natural insecticide)

Fig. 1.1

Narcotics and hallucinogens

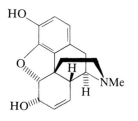

Morphine (opium)

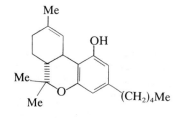

Tetrahydrocannabinol
(hashish, marijuana)

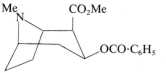

Cocaine (used primarily as a
stimulant by South American
Indians)

Stimulant

Caffeine

Perfumes and spices

Geraniol (rose oil)

Linalol (oil of lavender)

Fig. 1.1

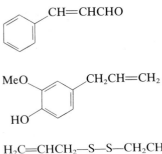

Cinnamaldehyde (cinnamon)

Eugenol (cloves)

$H_2C{=}CHCH_2{-}S{-}S{-}CH_2CH{=}CH_2$ Diallyl disulphide (garlic)

Fig. 1.1

component of curare, tubocurarine was the first drug used as a muscle relaxant in surgery. Ephedrine, the basis of an ancient Chinese remedy for respiratory ailments, was formerly used clinically as a drug for the treatment of asthma and hay fever. The psychoactive compounds morphine (opium) and cannabinoids (cannabis) have proved irresistible to mankind through many millennia. Caffeine was, and still is, the active principle of many native beverages. Crude extracts of bark, leaves, and seeds produce concoctions of considerable potency. In many ways our modern coffee, cocoa, tea, and cola are but poor imitations of these native brews. Finally, some of the odiferous and spicy compounds that have been used through the ages, such as linalol (oil of lavender), and eugenol (cloves), are also shown in Fig. 1.1.

In the light of the foregoing, it is not hard to understand what motivated the early nineteenth-century chemists in their efforts to isolate and characterize these natural products. Between 1815 and 1860 more than twenty of these active principles were isolated, including morphine, strychnine, quinine, caffeine, nicotine, codeine, camphor, and cocaine. However, accurate analyses were not possible before 1835, and even then it was rarely possible to do more than present molecular formulae and describe the characteristic reactions of the compounds. Many of these reactions were novel, and new ideas of molecular structure and reactivity followed. It was only natural that total syntheses of these compounds were then attempted, not only as a final confirmation of the structures, but because their structural complexity represented a considerable challenge to the synthetic chemist. Some of these compounds have defied the efforts of even the greatest chemists until quite recently (e.g. morphine, 1952).

It is worth noting that during this period of structure elucidation and total synthesis, the numerous new reactions that came to light excited much interest, and in several instances revolutionary new, and unifying concepts arose out of this work. Thus the studies of Barton on the reactions of the steroids (e.g. cholesterol and derivatives) led him to propound the principles of conformation and reactivity of cyclic systems (1956). Similarly, examination of the products of thermal and photochemical reactions of the vitamins D, and studies on the total synthesis of vitamin B_{12}, played some part in the process which culminated in the enunciation of the principles of conservation of orbital symmetry (Woodward and Hoffmann 1969).

As the structures of an increasing number of natural products became known, attempts were made to classify them in terms of structural type. This led quite naturally to speculation concerning their assembly or biogenesis. In some cases it was not difficult to spot characteristic structural features which suggested a particular progenitor or precursor. For example, many alkaloids incorporate the skeletons of simple amino acids, while terpenes and steroids usually contain an integral number of five-carbon units, originally thought to derive from isoprene (2-methylbuta-1,3-diene). This intuitive approach gave rise to certain unifying concepts such as 'the isoprene rule' (five-carbon precursor was isoprene or something rather similar), which was superseded by 'the biogenetic isoprene rule' (a biological equivalent of isoprene was employed as five-carbon progenitor, and the final structure could be rationalized in terms of 'chemically reasonable' modifications of a species formed from an integral number of C_5 units).

All of this was, of course, sheer speculation, but many of these early hypotheses have been shown to be amazingly accurate. In particular, many total syntheses of alkaloids were executed via hypothetical biosynthetic pathways before the actual biosynthetic pathways were established. The success of such 'biomimetic syntheses' provided the first real demonstration that biosynthesis might proceed via standard chemical reactions (albeit enzyme-catalysed) without recourse to 'biological trickery', i.e. there was no 'vital force'. In addition, the hypothetical intermediates on the biosynthetic pathways are often actual intermediates of the biomimetic sequence, or they occur in trace amounts in the same plant that contains the natural product under investigation. The probability that the hypothesis is close to reality is thus increased.

In the last twenty years, with the advent of nuclear magnetic resonance spectroscopy (n.m.r.), mass spectrometry, and routine X-ray crystallography, structure elucidation has become much more facile, and more time has been devoted to testing the biogenetic hypotheses. These investigations, in turn, have benefited enormously from the availability of precursor molecules, isotopically labelled with ^{14}C, ^{3}H, and most recently with ^{13}C, ^{2}H, ^{18}O, and ^{15}N. It is now possible in certain instances to determine the structure of a natural product, and establish the biosynthetic pathway, without recourse to any techniques other than ^{13}C-n.m.r. and the incorporation of ^{13}C-precursors. (An example of this is given in Chapter 2.)

Contemporary interest in natural products is thus turning increasingly to more biological topics: chemotaxonomy, enzyme studies, and chemical ecology. Chemotaxonomy is concerned with the description and classification of plants. If structurally similar compounds, derived from different plant species, are shown to share the same biosynthetic pathway, tentative assignment of both species to the same plant genus or family can be made. Of course many plants are already assigned to particular genera and families, but biogenetic relationships provide confirmation of the accuracy of these assignments, and in some instances demonstrate the need for reassignment.

Little is known about the enzymes of most biosynthetic pathways: their structure, mode of action, ease of inhibition, etc.; and the involvement of an increasing number of biochemists in such investigations is evidence that natural products are no longer solely of interest to chemists.

In many ways one of the most exciting things to emerge in the last few years is the realization that in many instances natural products do have a function in the organism from which they originate. Hitherto they have been considered to be detritus: useless but structurally interesting compounds. Now it is recognized that many of them have vital roles as mediators of ecological interactions; that is, they have a function in ensuring the continued survival of particular organisms, in an often hostile environment where many organisms are competing with each other: they thus increase the competitiveness of these organisms. Such ecological interactions will be considered in Chapter 7.

The study of natural products is now very much an interdisciplinary field, embracing chemistry and most of the biological subjects. However, lest we forget that the original *raison d'être* for these studies was the utilization of natural products in medicine, etc., it is worth noting that botanists now roam the less explored regions of Earth, seeking new and potentially useful plant species. In addition, many pharmaceutical companies are reinvestigating the properties of plants, long-used in native medicine, in the hope that novel medicinal agents may emerge. This strategy has already proved successful on numerous occasions, as for example with the podophyllins (Chapter 4)—a native cure for warts; and the vinca alkaloids (Chapter 6)—an alleged treatment for diabetes. In both cases new anti-cancer agents were discovered and are now in clinical use.

Primary and secondary metabolism

In the living organism (i.e. *in vivo*) chemical compounds are synthesized and degraded by means of a series of chemical reactions each mediated by an enzyme. These processes are known collectively as *metabolism*, comprising *catabolism* (degradation), and *anabolism* (synthesis). All organisms possess similar metabolic pathways by which they synthesize and utilize certain essential chemical species: sugars, amino acids, common fatty acids, nucleotides, and the polymers derived from them (polysaccharides, proteins, lipids, RNA, and DNA, etc.). This is *primary metabolism*, and these compounds, which are essential for the survival and well-being of the organism, are *primary metabolites*.

Most organisms also utilize other metabolic pathways, producing compounds which usually have no apparent utility: these are the 'natural products' we have been referring to in the preceding section. They are *secondary metabolites*, and the pathways of synthesis and utilization constitute *secondary metabolism*. These pathways are as much a product of the genetic make-up of the organism as are the primary pathways, but they are perhaps only activated during particular stages of growth and development, or during periods of stress caused by nutritional limitation or microbial attack.

The dividing line between primary and secondary metabolism is rather blurred: there are many obscure amino acids that are definitely secondary metabolites, while many steroid alcohols (sterols) have an essential structural role in most organisms and must therefore be considered as primary metabolites. In addition, the two types of metabolism are interconnected, since primary metabolism provides a number of small molecules which are employed as starting materials for all of the important secondary metabolic pathways. This is depicted in Fig. 1.2.

From this figure it will be apparent that there are three principal starting materials (or 'building blocks') for secondary metabolism:

(a) shikimic acid, the precursor of many aromatic compounds including the aromatic amino acids, cinnamic acids, and certain polyphenols;

(b) amino acids, leading to alkaloids, and peptide antibiotics including the penicillins and cephalosporins;

(c) acetate, precursor of polyacetylenes, prostaglandins, macrocyclic antibiotics, polyphenols, and the isoprenoids terpenes, steroids, and carotenoids), via two entirely separate biosynthetic pathways.

In each instance the precursor of these secondary metabolites is also used for the biosynthesis of certain classes of primary metabolites, e.g. proteins, fatty acids etc.

It is important to appreciate that most biological reactions are catalysed by enzymes, and are in principle reversible. In consequence we should spend a litle time considering enzymes and coenzymes, and the types of chemical reactions that they mediate.

Enzymes and coenzymes

Enzymes are proteins, that is poly-amino acids, which may have associated with them carbohydrates and cofactors (small organic molecules or metal ions) which are necessary for the structural integrity or catalytic activity of the enzyme. It used to be a common misconception amongst

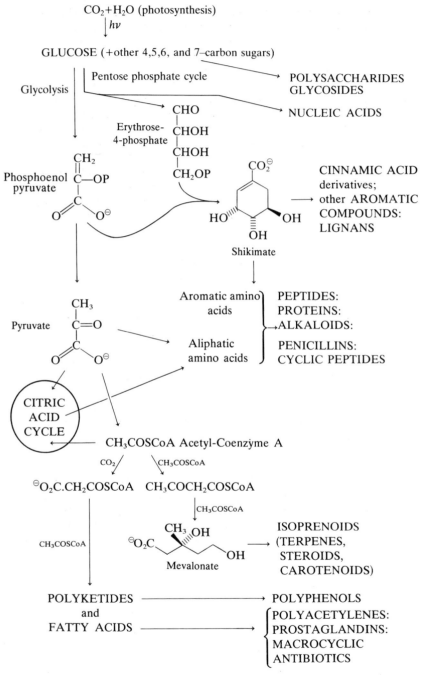

Fig. 1.2

chemists that enzymes are 'magic boxes' which bring about weird and wonderful transformations, unknown *in vitro*. In fact, with rare exceptions, enzymes catalyse the same types of reaction that are utilized in any organic chemistry laboratory: oxidation, reduction, alkylation, hydrolysis, hydroxylation, elimination, etc.

Enzymes do, however, enhance the rates of these reactions by as much as 10^9 to 10^{12}-fold. The reasons for this quite amazing catalytic behaviour are not really understood, but some general points may be stated.

(a) Usually each enzyme will catalyse only one particular type of reaction, and will only accept one substrate, or at the most a series of structurally similar substrates. This specificity means than an enzyme will almost invariably accept only one of the enantiomers present in any racemic substrate, and can thus produce an optically active product from an inactive substrate. Such specificity must be due to the three-dimensional arangement of substrate(s) and cofactor(s) at the active site of the enzyme; and futhermore it is probable that the substrate is 'stretched and strained' such that the enzyme–substrate complex which forms already has a transition-state configuration.

(b) Acidity and basicity are enhanced by the aprotic environment of the active site where solvent is excluded and acid–base catalysis is almost certainly involved in many enzymic reactions. This probably necessitates the transfer of a proton $(H^{\oplus})$ from neutral, or positively charged amino acid residues $(-SH, -OH, -NH_3^{\oplus}, -CO_2H)$ to other neutral, or negatively charged residues $(-S^{\ominus}, -NH_2, -CO_2^{\ominus})$.

(c) Finally, the sterospecific enzyme–substrate–cofactor complexes are stabilized primarily by non-covalent interactions. This means that they are readily dissociable (low or zero activation energy), and may dissociate spontaneously to release product(s) if the transition-state configuration is altered, resulting in a reduction in the number of stabilizing interactions.

Of the many cofactors employed by enzymes; three are of particular importance in all metabolic processes. These are: adenosine triphosphate (ATP), coenzyme A (CoASH), and $NAD(P)^{\oplus}/NAD(P)H$, the oxidized and reduced forms of nicotinamide adenine dinucleotide (phosphate). ATP and CoASH are 'reagents' that the enzyme employs to activate the substrate for subsequent reaction, and the redox couple $NAD(P)^{\oplus}/NAD(P)H$ is the mediator of enzymic oxidations and reductions. We shall discuss these three cofactors in turn.

ATP

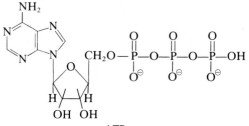

ATP

The general mode of action of ATP is shown in eqns (1.1) and (1.2).

$$\text{ROH} + \text{ATP} \xrightarrow[\text{enzyme}]{\text{Mg}^{2+}} \text{RO}-\overset{\displaystyle O}{\underset{\displaystyle O^{\ominus}}{\overset{\|}{\underset{|}{P}}}}-O^{\ominus} + \text{ADP} \qquad (1.1)$$

$$\text{RO}-\overset{\displaystyle O}{\underset{\displaystyle O^{\ominus}}{\overset{\|}{\underset{|}{P}}}}-O^{\ominus} + \text{HY} \xrightarrow[\text{enzyme}]{} \text{RY} + \text{HO}-\overset{\displaystyle O}{\underset{\displaystyle O^{\ominus}}{\overset{\|}{\underset{|}{P}}}}-O^{\ominus} \qquad (1.2)$$

The reaction almost invariably involves nucleophilic displacement of adenosine diphosphate (ADP) or of adenosine monophosphate (AMP) by a nucleophile $\text{RO}^{\ominus}$, to form a phosphate or pyrophsphate intermediate. A subsequent nucleophilic displacement of phosphate or pyrophosphate by another nucleophile, $Y^{\ominus}$, yields the product RY. It is well known from studies of phophate ester hydrolysis that the phosphate and pyrophosphate anions are good leaving groups, hence the second nucleophilic displacement is facilitated by formation of the phosphorylated intermediate. In case one should be tempted to think that such behaviour is unique to biological systems, it is only necessary to consider the uses of p-toluenesulphonyl chloride or of methanesulphonyl chloride in organic synthesis, to appreciate that the strategy is the same eqn (1.3).

$$\text{ROH} + \text{MeSO}_2\text{Cl} \rightarrow \text{ROSO}_2\text{Me} + \text{Cl}^{\ominus}$$

$$\text{ROSO}_2\text{Me} + Y^{\ominus} \rightarrow \text{RY} + \text{MeSO}_3^{\ominus} \qquad (1.3)$$

It is often erroneously stated that, *in vivo*, reactions are 'driven by the hydrolysis of ATP', i.e. an energetically unfavourable process, e.g. conversion of ROH into RY, eqn (1.4), is coupled to the hydrolysis of ATP, eqn (1.5), which is an exothermic process.

$$\text{ROH} + Y^{\ominus} + H^{\oplus} \leftrightharpoons R - Y + H_2O \qquad \Delta G^{\ominus} \approx + 20 \text{ kJ mol}^{-1} \ (1.4)$$

$$\text{ATP} + H_2O \leftrightharpoons \text{ADP} + \text{HOPO(OH)}_2 \qquad \Delta G^{\ominus} \approx - 31 \text{ kJ mol}^{-1} \ (1.5)$$

In fact, hydrolysis of ATP is the one reaction that must not take place if anabolic (i.e. biosynthetic) processes are to occur. Representation of phosphorylation as a two-step process as in eqns (1.1) and (1.2) (Y = PO_4^{3-} or $P_2O_7^{2-}$) is merely a form of 'energetic book-keeping', and has no chemical reality. It is much better to think in terms of the position of equilibrium of reversible reactions: $\Delta G^{\ominus}$ values are, of course, merely one way of expressing the values of equilibrium constants. Reaction mechanims are then most conveniently discussed by reference to the transient, reactive intermediates (phosphates, pyrophosphates, etc.) which are the favoured participants in these reversible processes.

CoASH

Coenzyme A is the most important acyl-transfer reagent in living organisms, and forms reactive acyl-thioester species, [1], on reaction with

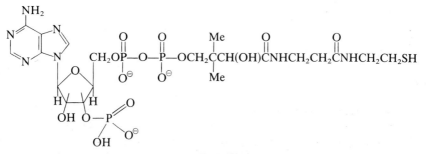

CoASH

acyl-substrates. The thioester group activates the acyl-species for nucleophilic attack at the carbonyl carbon atom, with displacement of CoAS$^\ominus$, and also for alkylation at the α-carbon atom (eqns 1.6 and 1.7).

$$RCH_2-CO-SCoA + YH \rightarrow RCH_2-CO-Y + CoASH \quad (1.6)$$
$$[1]$$
$$RCH_2-CO-SCoA + R'X \rightarrow RCHR' - CO - SCoA + HX \quad (1.7)$$
$$[1]$$

Again, we should not be surprised by such modes of reactivity: the ease of nucleophilic displacement in thioesters is well established, as is their ability to stabilize incipient carbanions, α to the carbonyl group. We shall discuss specific acyl-transfer reactions as they arise, but suffice to state at this point that such acylations and alkylations (eqns 1.6 and 1.7) are merely *in vivo* analogues of familiar *in vitro* reactions like the Claisen condensation and the numerous alkylations of activated carbonyl compounds. However, acyl-thioester species such as [1], have the advantage over many common organic compounds, in that they are bifunctional: they may act as electrophile (eqn 1.6) or nucleophile (eqn 1.7) with equal facility. The actual mode of reaction observed will depend upon the substrates and enzymes involved.

NAD(P)$^\oplus$/NAD(P)H

These cofactors are mediators of a large number of biological oxidations and reductions. In general, the enzymes of anabolism use the redox couple NAD(P)$^\oplus$/NAD(P)H for the mainly reductive process involved; and the enzymes of catabolism employ the couple NAD$^\oplus$/NADH for the mainly oxidative processes involved. When a substrate is oxidized, hydride (H$^\ominus$) is transferred from the substrate to C-4 of the nicotinamide ring of

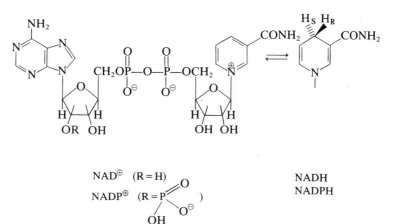

$$NAD^\oplus \quad (R = H)$$

$$NADP^\oplus \quad (R = P)$$

NADH
NADPH

$NAD(P)^\oplus$, thus producing $NAD(P)H$, and $H^\oplus$ is lost to the medium. The oxidation of ethanol to acetaldehyde is depicted in eqn (1.8), and as indicated, the process is reversible. Since the reduced form of the coenzymes has a prochiral centre (i.e. if one hydrogen is replaced by deuterium or tritium a chiral centre is produced), and many enzymes can differentiate between the hydrogens H_R, H_S, the process is usually stereospecific. In the example given in eqn (1.8), hydride from ethanol enters from above the plane of the ring, and it is this same hydrogen, H_R, which is transferred to acetaldehyde in the reverse process. Other systems are specific for the

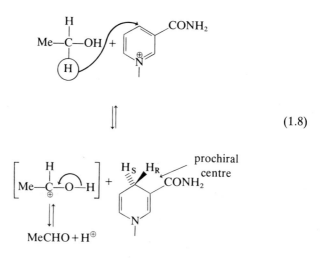

$$(1.8)$$

H_S hydrogen. In each instance, one $H^{\ominus}$ (of the cofactor) and $H^{\oplus}$ (medium) are utilized: this is depicted as 2[H] in subsequent figures.

The enzymes (dehydrogenases) catalysing these reversible hydride transfers have been shown to be folded into two distinct domains: the cofactor is bound in one, and the substrate in the other. By a sequence of reversible conformational changes, substrate and cofactor are presumably brought into a reactive configuration, and hydride transfer follows.

Transfer of hydride to organic molecules with concomitant reduction is a familiar *in vitro* process (LiAlH$_4$, NaBH$_4$, etc.), but the reverse process is rarely encountered. The closest analogy in organic chemistry is the Meerwein–Pondorf reaction, and its reverse, the Oppenauer oxidation: these are believed to proceed via hydride transfer.

Flavin coenzymes
The nicotinamide cofactors function as carriers of hydrogen, and are mobile in the sense that they can diffuse from enzyme to enzyme as depicted in eqn (1.9).

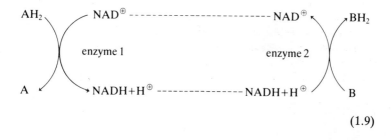

$$(1.9)$$

In contrast the flavin coenzymes, which also facilitate the transfer of hydrogen, are usually tightly bound to one particular enzyme. The mechanism of hydrogen transfer is not yet known, but one possible mechanism is shown in eqn (1.10). Reoxidation of reduced cofactor is normally effected directly or indirectly by molecular oxygen, and a possible free radical mechanism is depicted in Fig. 1.3. This involves one-electron transfers (rather than two as in eqn 1.9), and such a mechanism is necessitated because oxygen exists in its lowest energy state (ground state) as a triplet, i.e. as the diradical $\cdot O{-}O\cdot$. Addition of one electron yields the superoxide radical anion $\cdot O{-}O^{\ominus}$, while addition of a second electron produces the peroxide dianion $^{-}O{-}O^{-}$. Overall, the processes may be depicted as shown in eqn(1.11).

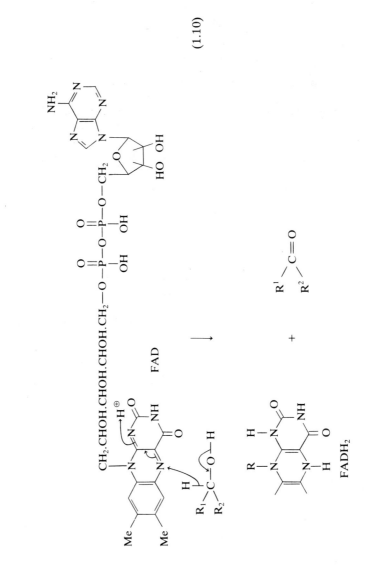

(1.10)

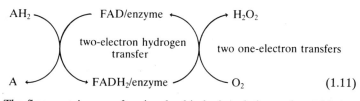

$$\text{(1.11)}$$

The flavoproteins are often involved in hydroxylation and epoxidation of organic substrates, and in other metabolic processes that involve active forms of oxygen (e.g. $O_2^{\ominus}$). However, these species are also produced by certain of the cytochromes (enzymes containing the cofactor haemin—see structure [48] Chapter 2), and here the actual chemistry involved is probably analogous to the *in vitro* chemistry depicted in eqns (1.12)—(1.15).

$$Fe^{2\oplus} + H_2O_2 \rightarrow Fe^{3\oplus} + HO^{\cdot} + HO^{\ominus} \qquad (1.12)$$

$$HO^{\cdot} + H_2O_2 \rightarrow HOO^{\cdot} + H_2O \qquad (1.13)$$

$$HOO^{\cdot} \leftrightharpoons H^{\oplus} + O_2^{\cdot\,\ominus} \qquad (1.14)$$

$$Fe^{3\oplus} + O_2^{\cdot\,\ominus} \rightarrow Fe^{2\oplus} + O_2 \qquad (1.15)$$

The enzymes concerned with these hydroxylations, epoxidations, etc., are usually known collectively as monooxygenases since they participate in the transfer of one atom from a molecule of oxygen, and they play a key role in the metabolism of drugs and other foreign substances.

Introduction of a one-carbon unit
Many secondary metabolites contain N-methyl, O-methyl, and aromatic C-methyl groups, and with few exceptions these methyl groups are derived from the cofactor, S-adenosyl methionine [2]. The probable course of reaction is depicted in eqn (1.16).

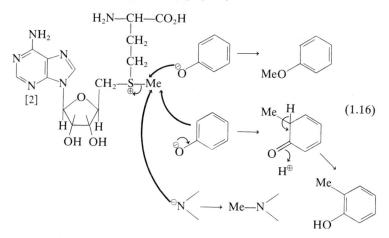

$$\text{(1.16)}$$

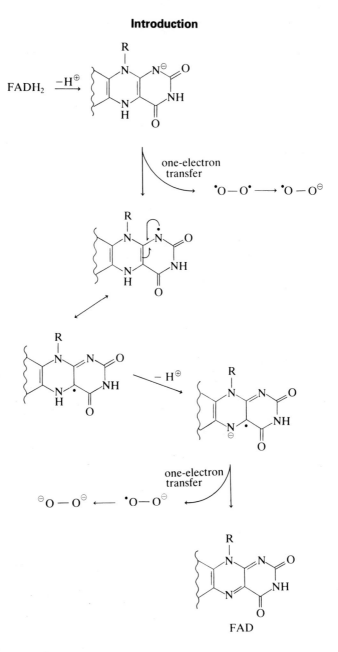

Fig. 1.3

The process is formally a nucleophilic displacement, and *in vitro* analogies are not hard to find, for example eqn (1.17).

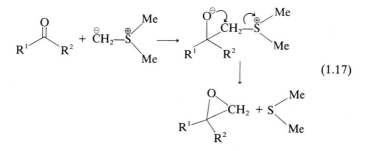

$$(1.17)$$

The actual source of the carbon atom that is transferred is variable, and it can be derived from formaldehyde, formic acid, ethanol, serine, etc. However, there is little doubt that S-adenosyl methionine is responsible for the actual transfer of the methyl group to the substrate, and recently good evidence for an S_N2-type transition state for this process has been obtained. Methionine, isotopically labelled with ^{14}C or ^{13}C has, in consequence, proved invaluable for establishing the origin of extraneous methyl groups in secondary metabolites.

This discussion of primary metabolism and of enzymes and coenzymes has been of necessity brief: for comprehensive coverage of these topics the reader is referred to the book in this series by Staunton (1978). Before leaving primary metabolism, it is worthwhile discussing three important primary processes that appear in Fig. 1.2: photosynthesis, carbohydrate metabolism, and the citric acid cycle, since these are responsible for the production of three of the principal building blocks of secondary metabolism.

Photosynthesis
All green plants, and certain algae and bacteria, have the capacity to utilize the electromagnetic energy of sunlight for the syntheses of adenosine triphosphate (ATP) and nicotine adenine dinucleotide phosphate (NADPH)

There are two primary light-mediated processes in photosynthesis:

(a) Absorption of light quanta by chlorophyll, or energy transfer to chlorophyll by other light absorbing pigments, causing electron transfer via a series of carrier species (a sequence of 'redox' (*reduction—oxidation*) reactions is involved). This is coupled to the production of NADPH and ATP.

(b) Photolysis of water to produce oxygen, and electrons which are similarly transferred via carrier species and coupled to production of NADPH and ATP.

In simple terms, the energy of sunlight is used in the synthesis of these two reactive species, which act as reducing and activating agent, respectively.

In a subsequent 'dark reaction', carbon dioxide is reduced to produce four-, five-, six-, and seven-carbon sugars: about 400×10^9 tonnes of carbon dioxide are fixed annually through the photosynthetic process. These sugars are metabolized via two major pathways: glucose catabolism (glycolysis) and the so-called pentose phosphate cycle. This latter process is in part identical to the basic pathway of sugar production and interconversion that comprises the 'dark reaction' of photosynthesis, but it does also occur in non-photosynthetic organisms. The major pathway of sugar metabolism is undoubtedly the ubiquitous process of glycolysis.

For a more detailed account of photoynthesis (and of glucose metabolism and the citric acid cycle), see Staunton (1978).

Glucose metabolism and the citric acid cycle
All cells have the capacity to metabolize glucose, by far the most important sugar, and in this way produce ATP, as well as small organic molecules which are the building blocks of the biosynthetic pathways. In the absence of oxygen, glucose is converted into pyruvic acid (Fig. 1.4), a process known as glycolysis; but in the presence of oxygen, glucose may be completely oxidized to carbon dioxide via the citric acid cycle, with concomitant production of many moles of ATP: this is known as respiration. The pathway of glycolysis is shown in Fig. 1.4, and it can be seen that a six-carbon molecule is converted into two three-carbon molecules (pyruvate). Two molar equivalents of ATP are consumed at the beginning of the sequence, but ATP is formed (via phosphorylation of ADP) concomitant with the formation of 3-phosphoglycerate and during the formation of pyruvate. The overall energy balance is thus a net gain of two molar equivalents of ATP, since two moles of pyruvate are produced for every mole of glucose consumed.

The product of this pathway, pyruvate, is converted into acetyl-S Coenzyme A, which is of pivotal importance for both primary and secondary metabolism. It may be used as a building block for the biosynthesis of fatty acids and derivatives, or for the biosynthesis of isoprenoids; but it may also enter the citric acid cycle. The products of the citric acid cycle are simple keto acids (which may react with ammonia or its equivalent to produce simple aliphatic amino acids), dicarboxylic acids, CO_2, and most important of all, ATP. The cycle is shown in Fig. 1.5. Acetyl-SCoA combines with oxaloacetate to produce citrate and the cycle then proceeds via the sequence shown to regenerate oxaloacetate, with formation of CO_2 and ATP, which is utilized in the biosynthetic processes. The overall stoichiometry of this process and associated processes is such that twelve molecules of ATP are produced for every molecule of acetate (as acetyl-SCoA), which is completely oxidized.

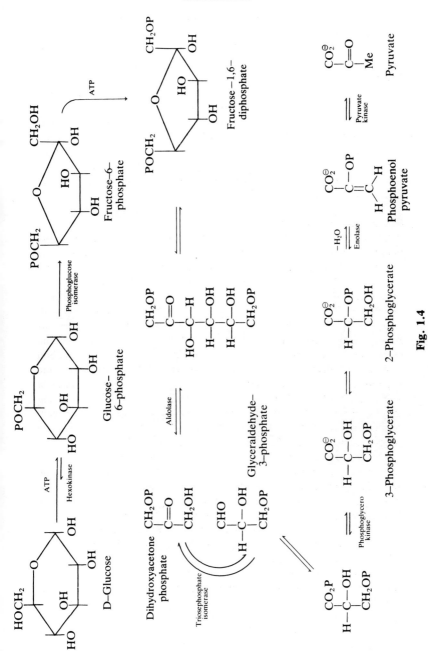

Fig. 1.4

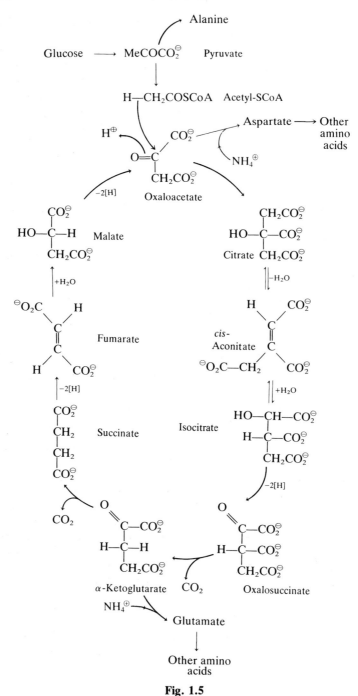

Fig. 1.5

Carbohydrates are thus the universal source of carbon atoms for metabolism, and provide the principal precursors for the biosynthesis of secondary metabolites: acetate, aliphatic amino acids (arising out of the citric acid cycle), shikimic acid (which is derived from phosphoenol pyruvate and erythrose-4-phosphate—see Fig. 1.2, p. 8), and aromatic amino acids (derived from shikimic acid). Photosynthetic organisms derive all of their organic compounds from carbon dioxide, but other organisms must obtain the basic building blocks of biosynthesis from the carbohydrates, amino acids, and fatty acids which are essential constituents of their diets. Despite these fundamental differences, the most important primary metabolic pathways differ only slightly from *Neurospora crassa* (bread mould) to *Homo sapiens* (Man): an excellent argument for the evolutionary process. The novelty lies with the secondary metabolic pathways, and we shall discuss these in the subsequent chapters.

Finally, it is readily apparent that the chemical species that lie at the end of the glycolytic pathway, i.e. phosphoenol pyruvate, pyruvate, and acetyl-SCoA, are those that are most important for secondary metabolism. In consequence, it has been suggested that any surplus of these compounds could be diverted (or 'siphoned off') into secondary metabolic pathways and away from the citric acid cycle. This would be a key feature of so-called 'shunt' or 'overflow' metabolism.

Elucidation of metabolic pathways

Although we shall be primarily concerned with proven pathways of secondary metabolism, it is worth considering how these routes have been established. After structure elucidation, it is usually possible to propose biogenesis from a particular precursor species. In the early stages this would be one of the major building blocks: acetate, mevalonate, shikimate, etc. Administration of isotopically labelled precursor to the organism producing the metabolite, is then tried. The metabolite is isolated, purified, and analysed for isotopic content, and if incorporation of isotope has occurred, cautious acceptance of the precursor/metabolite relationship is reasonable. It may then be possible to suggest a *biosynthetic pathway* from precursor to metabolite, and this may be tested by incorporation of specifically labelled (often stereospecifically labelled) precursors. It may also be possible to isolate proposed intermediates, which have incorporated label from the small precursor molecule, and which may then act as precursors for the metabolite under investigation (eqn 1.18).

$$A \rightarrow B \rightarrow C \rightarrow \cdots \rightarrow X \rightarrow Y \rightarrow Z \qquad (1.18)$$

A is a small, precursor species. When administered, isotopically labelled, B, C, etc. to X and Y may (in theory) be labelled in turn, and used as precursors of metabolite Z. The use of mutant organisms can often provide additional information about these pathways. The mutation is usually the result of damage caused by X-rays, UV, or certain chemicals acting upon the genes (i.e. chromosomal DNA), with resultant aberrant enzyme production. Thus, for example, the enzyme catalysing the conversion B in eqn (1.18) may be absent or defective, and metabolite B should then accumulate, though it may also be diverted into new or separate pathways. This is the ideal result, but a lethal mutation is much more likely, and here primary and secondary metabolism is so defective that life is not possible. In this way, biogenetic hypotheses become biosynthetic demonstrations.

Two difficulties confront the investigator:

(a) incorporation of a sufficiently high percentage of labelled precursor to render the results meaningful; and

(b) the need to analyse (and perhaps degrade) the labelled metabolite in order to establish which atoms have been labelled.

Administration of precursor

Labelled precursors may be administered to intact organisms or to cell-free extracts of them. In general, plants incorporate label rather poorly (often as little as 10^{-2} to 10^{-4} per cent of the total fed, possibly because administered precursor is unable to penetrate to the site of biosynthesis, or because it is degraded *en route*. Better results are obtained with cell-free extracts from animals, and with bacterial and fungal cultures (usually > 1 per cent incorporation, and sometimes >10 per cent). Many investigators have thus turned their attentions to the use of cell-free systems (i.e. a crude mixture of enzymes from the organism) or to tissue cultures. The latter are usually obtained from small pieces of plant tissue grown in such a way that they produce undifferentiated callus cells, which may retain the capability to produce certain secondary metabolites, but have lost the capability to produce others. One eventual commercial aim of this technique is to provide cell lines, which only produce certain alkaloids or terpenoids for medicinal or perfumery purposes. An extension of this aim is the production of clones of the genes that govern the biosynthesis of secondary metabolites. This should not only allow large-scale production of selected secondary metabolites, but also offers the intriguing possibility of synthesizing completely novel metabolites through the use of modified gene clones (i.e. through genetic engineering). Further discussion of such methodology is beyond the scope of this book, but the interested reader is referred to the review by Hutchinson (1986) for further information and leading references.

In addition, it is important to realize that administration of what often amounts to large quantities of somewhat artificial substrates, may disturb the *status quo* of the organism. Further, since most secondary metabolites are isolated through destruction of the organism, we can never be absolutely certain that they are true metabolites, and not the products of a *post mortem* chemical change. For these reasons, results obtained in biosynthetic experiments must always be interpreted with caution.

Examination of labelled metabolites

Before analysis commences, a pure sample of isotopically labelled compound must be available. Many erroneous results have been obtained when supposedly pure metabolites have been contaminated with minute amounts of, for example, highly radioactive impurities. The commonly employed radioisotopes are tritium (^3H), a β-emitter with half-life of 12.1 years and carbon-14 (^{14}C), a β-emitter with half-life 5640 years. The radiolabelled metabolites are usually degraded to identify the centres of enrichment.

All of the common degradative reactions of organic chemistry have been employed: ozonolysis, decarboxylation, thermal and photochemical bond cleavage, etc. However, each degradative step reduces the amount of labelled material available for subsequent steps, and it is often impossible to complete a degradative sequence. In consequence, it is difficult to obtain a complete labelling pattern when ^3H and ^{14}C are used as isotopic labels, and it is for this reason that the use of the isotope ^{13}C has enormous potential. The isotope has a natural abundance of 1.1 per cent, and a nuclear spin $I = \frac{1}{2}$, hence n.m.r. spectra may be obtained given suitable instrumentation. Degradation of the labelled compound is no longer necessary since carbon atoms which are isotopically enriched can be located by a comparison of the enriched spectrum with the natural abundance spectrum.

Deuterium (^2H)-n.m.r. has also gained prominence in recent years for the same reasons, and both techniques will be discussed further in Chapter 2.

Little more will be said concerning actual biosynthetic experimentation: we shall be primarily concerned with the results obtained. However, it is hoped that, as we pass on to consider those secondary metabolic pathways that have been elucidated, the reader will bear in mind the experimental difficulties just discussed.

2. Secondary metabolites derived from acetate: fatty acids and polyketides

A very large number of natural products are derived from acetyl-coenzyme A; one of the basic building blocks for secondary metabolism, which is itself formed from acetate. A summary of the metabolic pathways which commence with acetyl-coenzyme A is given in Fig. 2.1.

It is evident that there are two major pathways: one which proceeds by the stepwise addition of C_2 units, and the other by condensation of C_5 units to produce isoprenoids. We shall defer discussion of this latter pathway until Chapter 3.

In this chapter we shall consider the assembly of the linear 'polyketide' chain, and its subsequent modification (reduction, cyclization, etc.) to produce either fatty acids and derivatives, or polyketides these are sometimes called, collectively, the _acetogenins_. It is convenient to study the fatty acids first, since, structurally, they are the simplest metabolites of acetate.

Saturated fatty acids

The most abundant saturated fatty acids, of general formula $MeCH_2(CH_2CH_2)_nCH_2CO_2H$, are:

caprylic-C_8; $n = 2$	lauric-C_{12}; $n = 4$	palmitic-C_{16}; $n = 6$
capric-C_{10}; $n = 3$	myristic-C_{14}; $n = 5$	stearic-C_{18}; $n = 7$

They occur _in vivo_ as constituents of natural waxes and seed oils; in glycerides, that is, esters formed from glycerol and one, two, or three fatty acids; and in phospholipids, as for example in phosphatidyl choline [3].

$$
\begin{array}{l}
\mathrm{CH_2OCOR^1} \\
\mathrm{R^2COOCH \quad O^{\ominus}} \\
\mathrm{CH_2OPOCH_2CH_2\overset{\oplus}{N}Me_3} \\
\mathrm{\qquad \underset{O}{\|}}
\end{array}
\qquad [3]
$$

Such polar lipids are of vital importance for the structural integrity of cell membranes, which control such properties as contractility and permeability. _always present._

The ubiquity of many of the common fatty acids, and the vital roles many of them have, puts them into the class of primary metabolites; and it is

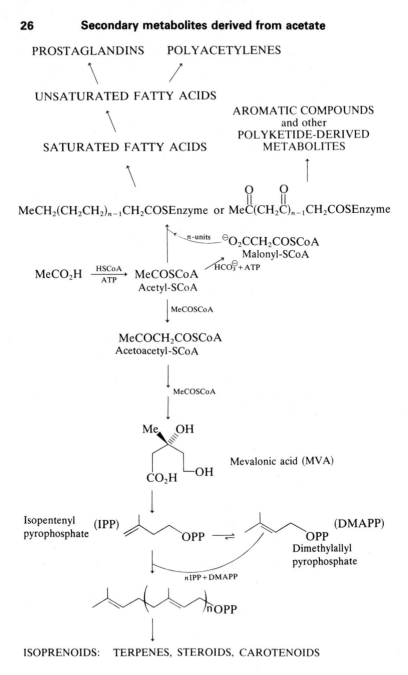

Fig. 2.1

only the more unusual or uncommon fatty acids that can be considered as true secondary metabolites.

Fatty acids with chain length greater than C_{20} are rare, but do occur in natural waxes, usually as esters with sterols (steroid alcohols), or with long-chain aliphatic alcohols (C_{16} to C_{36}, often derived from fatty acids). Beeswax for example contains palmitic and cerotic (C_{26}) acids esterified with melissyl alcohol (C_{30}). The lower acids occur mainly in animal fats: cow's milk fat contains considerable amounts of butanoic acid (C_4) together with lesser amounts of C_6, C_8, C_{10} and C_{12} acids.

In our modern society fatty acids have an important role as major components of soaps and detergents. Sodium salts of lauric and myristic acids are used in domestic soaps, while the potassium salts are employed in more specialized preparations such as shaving cream and liquid soap. Natural detergency is first apparent with lauric acid, but many synthetic detergents are now employed instead of these natural 'soaps'. An important example is sodium glyceryl monolaurate sulphate [4]:

$$CH_2OCO(CH_2)_{10}Me$$
$$|$$
$$CHOH \qquad\qquad [4]$$
$$|$$
$$CH_2OSO_2ONa$$

These synthetic detergents owe their usefulness to an enhanced resistance to hard water.

Finally, it should be noted that consumption of food with a high content of saturated fatty acids (or fats derived from them) is supposed to predispose the consumer to coronary artery disease. Adherents of this belief advocate the consumption of food containing vegetable fats (high in unsaturated fatty acids) rather than food containing animal fats (high in saturated fats). The efficacy of such a diet in the prevention of arteriosclerosis in Man remains to be fully substantiated.

Biosynthesis
The inherently simplest mode of biosynthesis would be a linear polymerization of acetyl-units. (CH_3CO-). Chemically this is quite acceptable: we have already seen that the thio-coenzyme A group (SCoA) will stabilize anions α to the carbonyl group of thiol esters, and is in addition a good leaving group. Hence a series of Claisen condensations could conceivably produce a linear polyketide, as shown in eqn (2.1). Reduction of this polyketide chain should then produce fatty acids.

Indeed, this mode of biosynthesis is known *in vivo*, but occurs primarily in organisms where oxidative degradation (so-called β-oxidation, and formally the reverse of the process depicted in eqn 2.1) provides a large 'pool' of acetyl-SCoA. It is also the route by which acetoacetyl-SCoA [5] is produced, and this intermediate is on the biosynthetic pathway that leads to the isoprenoids.

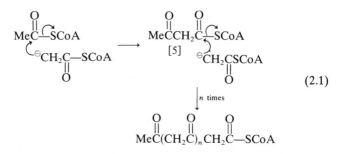

$$(2.1)$$

The more common biosynthetic pathway was elucidated following the discovery that carbon dioxide (from bicarbonate) was necessary for *de novo* biosynthesis, and further, that label from [^{14}C]-bicarbonate was not incorporated into the newly synthesized fatty acids. These somewhat surprising results were explained when malonyl-SCoA [6] was shown to be the source of C_2 units which were added to acetyl-SCoA, as starter unit. Malonyl-SCoA is likely to be a better nucleophile than acetyl-SCoA, especially if a concerted loss of carbon dioxide accompanies attack at the carbonyl group of acetyl-SCoA (eqn 2.2). (In order to simplify mechanistic equations, the presumed tetrahedral intermediates will be left out of subsequent equations.)

$$(2.2)$$

Malonyl-SCoA itself is formed from bicarbonate and acetyl-SCoA via the reaction sequence depicted in eqn (2.3).

$$ATP + HCO_3^{\ominus} + \text{Biotin carboxyl carrier protein} \rightleftharpoons$$
$$\text{(BCCP)}$$
$$^{\ominus}O_2C - BCCP + ADP + P_i \qquad (2.3)$$

$$^{\ominus}O_2C - BCCP + MeCOSCoA \rightleftharpoons BCCP + ^{\ominus}O_2CCH_2COSCoA$$
$$[6]$$

The cofactor biotin, [7] (R = OH; R$'$ = H), is involved in this process and the CO_2 moiety is transferred to acetyl-SCoA via a 1$'$-*N*-carboxy biotin-enzyme intermediate [7] (R = enzyme; R$'$ = CO$_2^{\ominus}$). It is now apparent why carbon dioxide is necessary for *de novo* biosynthesis, and why label from [^{14}C]-bicarbonate is not incorporated into the fatty acids produced.

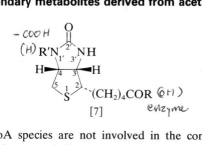

[7]

In fact thio-CoA species are not involved in the condensation step eqn (2.2), but analogous acyl-thiol ester intermediates are used. The most important of these are the acyl-derivatives of acyl carrier protein (ACP). This possesses a phosphopantetheine side-chain (the same side-chain as in co-enzyme A, from whence it derives) to which the various acyl-groups are joined as thiol esters [8].

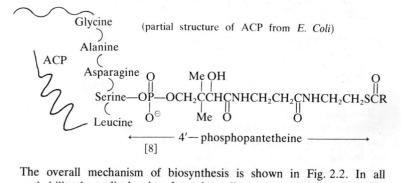

The overall mechanism of biosynthesis is shown in Fig. 2.2. In all probability the cyclical series of reactions all take place within the confines of a multi-enzyme complex, with individual thiol ester intermediates passing from one enzyme thiol group to the next. In step 1, the acetyl group of acetyl-SCoA is transferred to the acyl carrier protein, and in step 2 it is subsequently transferred to the active site thiol of the condensing enzyme. Malonyl-SCoA is similarly transformed into malonyl-SACP (step 3), and condensation follows (step 4). Stereospecific reduction mediated by NADPH (step 5) produces the (3R)-hydroxy intermediate, exclusively, and elimination (step 6) produces a 2-(E)-enoyl-SACP species. The cycle is completed by further reduction mediated by NADPH, to produce a saturated acyl-SACP intermediate (step 7).

The stereochemical aspects of each step have been probed, and it is known from studies with chiral acetates (i.e. acetate which possesses one atom of each of the isotopes of hydrogen, and of known configuration) that the carboxylation step occurs with retention of configuration, whilst the condensation with malonyl-SCoA occurs with inversion of configuration:

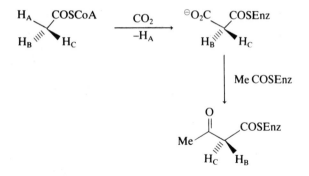

The elimination reaction (step 6) has been shown to proceed with overall *syn* stereochemistry by using the modified substrate shown in Fig. 2.2.; and the final reduction (step 7) involves *syn-* or *anti-*addition of the elements of hydrogen ($H^{\ominus}$ and $H^{\oplus}$) depending upon the organism concerned. Face selectivity is also observed (Fig. 2.2). Repetition of this cycle (steps 2 to 7 inclusive) utilizing the newly formed acyl intermediate in place of acetyl-SCoA, leads to lengthening of the carbon chain by two carbon atoms every cycle, to yield acyl-species of the general form $MeCH_2(CH_2CH_2)_nCH_2COSACP$.

This process terminates when the chain length reaches C_{16} or C_{18}, yielding palmitic or stearic acids, or their thiol esters. It is probable that as the chain length approaches C_{16}–C_{18}, the active-site thiol of the condensing enzyme has a greater affinity for an acetyl-SACP species. That is, steric or electronic effects hinder access of acyl substrates bigger than C_{16}–C_{18} to the active site, and termination of chain extension results.

Extension of the chain beyond C_{18} does occur, and both acetyl-SCoA and malonyl-SCoA serve as donors of C_2 units; which of the two is employed varies from species to species, and even from one part of an organism to another. The ultimate chain length is rarely greater than C_{22}–C_{24}, except in higher plants, where chain lengths of up to C_{30} are encountered: many natural plant waxes contain C_{29} paraffins formed by decarboxylation of a C_{30} fatty acid.

Finally, it should be noted that other acyl-SCoA moieties may function as 'starter' units, in place of the usual starter acetyl-SCoA. The same synthetase complex can be used, and if propionyl-SCoA is added, a mixture of odd chain length fatty acids is produced, e.g. C_{15}: C_{17} (1:3), in a typical experiment, while butyryl-SCoA yielded C_{14}:C_{18} (1:9). In each case seven moles of malonyl-SCoA are utilized to produce the predominant products. It would appear that it is the availability of aberrant starters that detemines whether normal or abnormal fatty acids are produced, rather than a requirement for special enzymes able to cope with them.

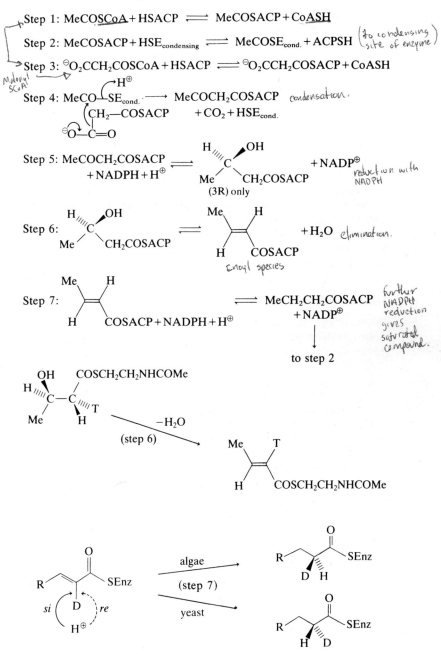

Step 1: MeCOSCoA + HSACP $\rightleftharpoons$ MeCOSACP + CoASH

Step 2: MeCOSACP + HSE$_{condensing}$ $\rightleftharpoons$ MeCOSE$_{cond.}$ + ACPSH $\left(\begin{array}{l}\text{to condensing}\\\text{site of enzyme}\end{array}\right)$

Step 3: $^{\ominus}O_2CCH_2COSCoA$ + HSACP $\rightleftharpoons$ $^{\ominus}O_2CCH_2COSACP$ + CoASH *Malonyl SCoA*

Step 4: MeCO$-$SE$_{cond.}$ $\xrightarrow{H^{\oplus}}$ MeCOCH$_2$COSACP *condensation.*
$\overset{\oplus}{(}$CH$_2$$-$COSACP + CO$_2$ + HSE$_{cond.}$
$^{\ominus}O-C=O$

Step 5: MeCOCH$_2$COSACP $\rightleftharpoons$ [structure (3R) only] + NADP$^{\oplus}$ *reduction with NADPH*
    + NADPH + H$^{\oplus}$

Step 6: [structure] $\rightleftharpoons$ [Enoyl species] + H$_2$O *elimination.*

Step 7: [structure] COSACP + NADPH + H$^{\oplus}$ $\rightleftharpoons$ MeCH$_2$CH$_2$COSACP + NADP$^{\oplus}$ *further NADPH reduction gives saturated compound.*

to step 2

[structure with T] $\xrightarrow[\text{(step 6)}]{-H_2O}$ [structure]

[structure with si/re, D, H$^{\oplus}$] $\xrightarrow[\text{(step 7)}]{\text{algae}}$ [structure]
$\xrightarrow{\text{yeast}}$ [structure]

Fig. 2.2

A good example of this phenomenon is provided by the thermophilic organism, *Bacillus acidocaldarius*, which inhabits the hot water pools associated with Vesuvius. It grows best at 60°C, and synthesizes a variety of linear and cyclic fatty acids, including several derived from cyclohexane carboxy-SCoA [9]. (This is in fact derived from shikimate, rather than

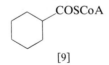

[9]

from acetate.) The synthetase of this organism will also accept a wide variety of artificial starters, including: straight chain units of five, six, or seven carbon atoms, branched chain starters, and cyclic acyl-species derived from cyclobutyl acetic acid, cyclopentyl acetic acid, and cycloheptyl acetic acid. It is perhaps hardly suprising that an organism which chooses to inhabit such a strange and alien place, should possess an enzyme system which is equally unusual.

Unsaturated fatty acids

The majority of naturally occurring unsaturated fatty acids are in the C_{18} series. Acids shorter than C_{14}, or longer than C_{22}, are rare. Some representative examples are:

palmitoleic (C_{16})

which occurs in nearly all fats, especially those of marine origin;

oleic (C_{18})

which represents 83 per cent of total fatty acids present in olive oil, and 60 per cent of total in peanut oil;

linoleic (C_{18})

representing 21 per cent of fatty acids in peanut oil;

α-linolenic (C_{18})

(γ-linolenic is the 6, 9, 12-triene);

ricinoleic (C_{18})

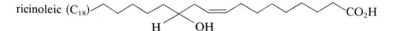

representing about 88 per cent of the total fatty acids of castor oil; arachidonic (C_{20})

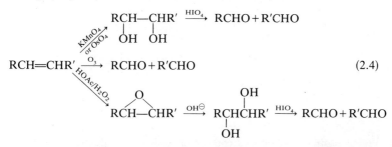

which is found in mammalian adrenal glands.

With rare exceptions, the double bonds in mono- and polyunsaturated fatty acids are *cis* or *Z*, and this fact must be explained by any biogenetic hypothesis.

The chemistry of these acids is not remarkable: it is largely that expected of mono- and non-conjugated, polyolefinic systems. Their structures have been determined by classical oxidative methods, as shown in eqn (2.4):

$$
\begin{array}{c}
RCH{-}CHR' \xrightarrow{HIO_4} RCHO + R'CHO \\
\underset{KMnO_4 \text{ or } OsO_4}{\nearrow} \; \overset{|}{OH} \; \overset{|}{OH}
\end{array}
$$

$$RCH{=}CHR' \xrightarrow[HOAc/H_2O_2]{O_3} RCHO + R'CHO \qquad (2.4)$$

$$
\underset{RCH{-}CHR'}{\overset{O}{\triangle}} \xrightarrow{OH^{\ominus}} \underset{\overset{|}{OH}}{RCHCHR'} \xrightarrow{HIO_4} RCHO + R'CHO
$$

Commercially, the most important class of reactions are the autoxidations, since autoxidation of the polyunsaturated acids in margarine leads to rancidity; and thus antioxidants must be added. A typical autoxidation of linoleic acid is shown in eqn (2.5)

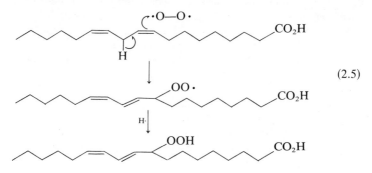

$$(2.5)$$

Biosynthesis

Two pathways to unsaturated fatty acids have been delineated: one, in which there is an absolute requirement for oxygen (aerobic route), and the other, which may proceed in the absence of oxygen (anaerobic route). The former process is by far the most common, and operates in yeasts, certain bacteria, algae, higher plants, and vertebrates. The anaerobic pathway is mainly confined to the anaerobic bacteria.

Aerobic route. Although certain minor variations are encountered between organisms, the basic progress involves successive desaturations (eqn 2.6). Thus linoleic-thiol ester is formed from stearyl-SACP. In

$$Me(CH_2)_{16}COSX$$

$$\downarrow O_2/NADPH$$

$$\text{(2.6)}$$

$$Me(CH_2)_7CH\overset{Z}{=}CH(CH_2)_7COSX$$

$$\downarrow O_2/NADPH$$

$$Me(CH_2)_4CH\overset{Z}{=}CHCH_2CH\overset{Z}{=}CH(CH_2)_7COSX$$

virtually all systems, palmityl- and stearyl-thiol esters are the preferred substrates of the various desaturase enzymes. In higher plants the thiol ACP species appear to be the actual substrates, while in yeasts and animals the thiol CoA species are involved.

The positions of the second and subsequent desaturations are non-random. In most, non-mammalian systems, new double bonds are introduced, progressively towards the methyl terminus. Animals have lost the capacity to desaturate at C-12 or at any position between an extant double bond and the methyl terminus, and desaturation proceeds towards the carboxy terminus with concomitant chain elongation. These pathways are shown in Fig. 2.3.

A consequence of this metabolic block in animals, is that linoleic acid must be a constituent of their diets, since essential tetraenoic acids, such as arachidonic acid, which have a double bond beyond C-12, can only be synthesized from a C_{18} precursor which already has a double bond 'to the left of' the original point of unsaturation (Fig. 2.3).

The mechanism of the desaturation process is at present unknown, but it appears to proceed via a concerted *syn*-elimination of a pair of hydrogens (vicinal-pro-R-hydrogens), without the intermediacy of any oxygenated species. A plausible scheme has been suggested by James and Gurr (eqn 2.7).

Finally, it should be noted that it is still uncertain whether higher plants synthesize their major monoenoic acids (C_{16} and C_{18}) by chain elongation of a C_{12} or C_{14} monoenoic acid, as well as by desaturation of palmityl- or stearyl-thiol esters. There is little evidence for or against this mechanism, but the process does require oxygen, unlike this anaerobic route, which is otherwise rather similar.

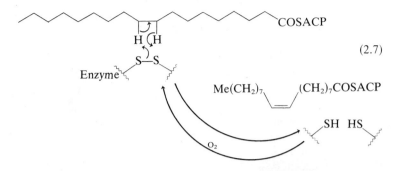

(2.7)

Anaerobic route. That unsaturated fatty acids can be synthesized in the absence of oxygen is apparent from their widespread occurrence in anaerobic bacteria. However, only monoenoic acids are synthesized, and the following facts have been established in radiochemical experiments:

(a) [1-^{14}C]-octanoic acid gave Δ^9-C_{16} and Δ^{11}-C_{18} acids, which contained isotopic label, as well as labelled saturated acids.

Me(CH$_2$)$_7$CH=CH(CH$_2$)$_7$COSX $\longrightarrow$
    Oleate

$\qquad$ Me(CH$_2$)$_4$CH=CHCH$_2$CH=CH(CH$_2$)$_7$—COSX
    $\qquad\qquad$ Linoleate

animals $\qquad\qquad$ plants

$\qquad$ MeCH$_2$CH=CHCH$_2$CH=CHCH$_2$CH=CH(CH$_2$)$_7$—COSX
    $\qquad\qquad\qquad\alpha$-Linolenate
    $\qquad\qquad$ (all double bonds are *cis*-Z)

Me(CH$_2$)$_7$CH=CHCH$_2$CH=CH(CH$_2$)$_4$COSX

$+C_2$

Me(CH$_2$)$_7$CH=$\overset{11}{C}$HCH$_2$CH=$\overset{8}{C}$HCH$_2$CH=$\overset{5}{C}$H(CH$_2$)$_3$COSX
    Eicosatri-5,8,11-enoate

$\qquad$ Me(CH$_2$)$_4$CH=CHCH$_2$CH=CH (CH$_2$)$_7$COSX
    $\qquad\qquad$ (dietary) linoleate

$+C_2$
desaturation

Me(CH$_2$)$_4$CH=$\overset{14}{C}$H—CH$_2$CH=$\overset{11}{C}$H—CH$_2$CH=$\overset{8}{C}$HCH$_2$CH=$\overset{5}{C}$H(CH$_2$)$_3$—COSX
    Eicosatetra 5,8,11,14-enoate (arachidonic acid)

Fig. 2.3

(b) $[1\text{-}^{14}C]$-decanoic acid gave $\Delta^7\text{-}C_{16}$ and $\Delta^9\text{-}C_{18}$ acids and saturated acids, all of which were labelled.

(c) $[1\text{-}^{14}C]$-dodecanoic acid and $[1\text{-}^{14}C]$-tetradecanoic acid gave only labelled saturated fatty acids.

Hence it seems likely that a branching point occurs at the C_8 or C_{10} stage, and the mechanism is believed to be as shown in Fig. 2.4.

The key step is catalysed by an enzyme which functions as both dehydratase and isomerase, producing $2(E)$- and $3(Z)$-decenoates. Since only the E-isomer is a substrate in the second reduction of the cycle by which saturated acids are synthesized, the Z-double bond is retained through subsequent cycles. In this way both saturated and Z-monoenoic acids are produced. A 3-decynoyl-thioester specifically inhibits biosynthesis of *cis*-acids, presumably by interference with the isomerization step. Elegant experiments, using (4R)- and (4S)-specifically deuterated dec-2-enoic acids in conjunction with a mutant strain of *E. coli*, established that the (pro-4R)-hydrogen was lost in the

Me(CH₂)₅CH₂COSX
Octanoate

$+C_2$ and reduction

Me(CH₂)₅CH₂CH(OH)CH₂COSX

dehydratase

Me(CH₂)₅CH$\overset{Z}{=\!=\!=}$CHCH₂COSX ⟵ isomerase ⟶ Me(CH₂)₅CH₂CH$\overset{E}{=\!=\!=}$CHCOSX
3-(Z)-decenoate   (inhibited by Me(CH₂)₅C≡CCH₂COSX)   2-(E)-decenoate

NADPH

Me(CH₂)₅CH₂CH₂CH₂COSX

$+4C_2$

$+4C_2$

Me(CH₂)₅CH$\overset{Z}{=\!=\!=}$CH(CH₂)₉CO₂H
Vaccenic acid

Me(CH₂)₁₆CO₂H
Stearic acid

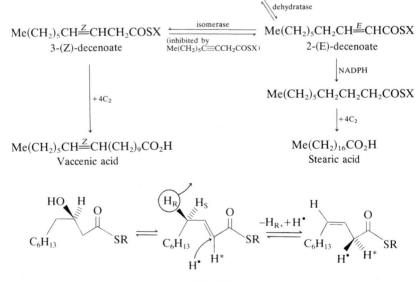

Fig. 2.4

course of the rearrangement reaction. In addition, by using [2-²H] dec-2-enoic acid, it was possible to demonstrate that the incoming hydrogen was added from the *si*-face (Fig. 2.4).

Most of the fatty acids discussed thus far are of rather widespread occurrence, and do not posssess the novel structures characteristic of most secondary metabolites. There are, however, many naturally occurring fatty acid metabolites of very restricted occurrence, and we should consider a few examples.

The flavour of Bartlett pears is due at least in part to the esters [10], while the aroma and flavour of black tea leaves derives from the 68 volatile constituents that have been identified: these include many simple

$$\text{Me(CH}_2)_4\text{CH}\overset{Z}{=}\text{CHCH}_2\text{CH}_2\text{CH}\overset{E}{=}\text{CHCO}_2\text{R}$$

(R = Me or Et)

[10]

aliphatic compounds derived from fatty acids, as well as from amino acids, and isoprenoids. Many insects utilize aliphatic esters, alcohols, and even alkanes, as sex attractants, alarm substances, and population-controlling agents. Thus [11] is the principal sex attractant of the silkworm moth, [12] is an alarm substance released by the common bed-bug, which serves to warn other members of the species of impending danger, and [13] is an 'overcrowding' factor produced by mosquito larvae when population densities reach critical levels, and which acts by inhibiting the growth of younger larvae.

$$\text{HOCH}_2(\text{CH}_2)_8\text{CH}\overset{E}{=}\text{CHCH}\overset{Z}{=}\text{CHCH}_2\text{CH}_2\text{Me} \qquad [11]$$

$$\text{MeCH}_2\text{CH}_2\text{CH}\overset{E}{=}\text{CHCHO} \qquad [12]$$

$$\begin{array}{c}\text{Me}\\ \diagdown\\ \text{CH(CH}_2)_5\text{Me} \qquad [13]\\ \diagup\\ \text{Me}\end{array}$$

Finally the cyclic species [14] is released from the female gametes of the marine alga, *Ectocarpus siliculosus*, and serves as an attractant for the

[14]

male reproductive organelles. Many more such compounds will be mentioned in the final chapter, when we consider the ecological significance of secondary metabolites.

We now pass on to consider two groups of metabolites which are the result of extensive structural modification of fatty acid precursors: the polyacetylenes, and prostanoids and leukotrienes.

The polyacetylenes

The first natural acetylene to be discovered (1892) was tariric acid [15]. Progress in discovering other natural acetylenes was slow, and as recently as 1950 only a dozen or so were known. Since then, however, due mainly to the efforts of E. R. H. Jones, and Bohlmann, the list of known compounds is around one thousand. They have characteristic UV absorption spectra and are easily detected now that sensitive instruments are available.

$$Me(CH_2)_{10}C{\equiv}C(CH_2)_4CO_2H \qquad\qquad [15]$$

They are widely dispersed in nature, and are especially common in plants of the families Compositae (daisy), and Umbelliferae (e.g. parsley and carrot), as well as in many fungi. An almost bewildering variety of structural types are encountered. Daisy seeds contain crepenynic acid [16] and dehydromatricaria ester [17]: the common names are as exotic as the structures. *Dahlia coccinea* contains large amounts of the tetrahydropyranyl acetate [18], while Compositae are a rich source of acetylenic thiophenes, such as [19]. Finally, the somewhat bizarre, antiviral acetylene, chondriol [20], has recently been isolated from a marine alga.

$$Me(CH_2)_4{-}C{\equiv}C{-}CH_2CH\overset{Z}{=}CH(CH_2)_7CO_2H \qquad [16]$$

$$Me{-}C{\equiv}C{-}C{\equiv}C{-}C{\equiv}C{-}CH\overset{E}{=}CHCO_2Me \qquad [17]$$

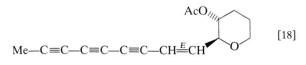

[18]

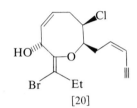

Me—C≡C—$\langle$ S $\rangle$—CH$\overset{E}{=}$CHCO₂Me   [19]

[20]

Very little is known concerning the biological function of these metabolites. It has been suggested that they serve as substrates for other metabolic processes, and this is substantiated in part by the finding that polyacetylenes can be isolated on the day a seed begins to germinate. Many of the compounds are extremely active antibiotics, but their toxicity precludes any therapeutic application. The compound falcarinone [21] occurs in carrots and is acutely toxic, but is present in such small amounts that it is not dangerous to the consumer. Other acetylenes have anti-fungal properties, and in at least two cases have been shown to be produced by the plant in response to the ravages of an invading organism, but not otherwise. One such example, is wyerone [22], produced by the broad-bean plant in response to an invasive pathogen.

$$CH_2{=}CH{-}COC{\equiv}C{-}C{\equiv}C{-}CH_2CH\overset{Z}{=}CH(CH_2)_6Me \qquad [21]$$

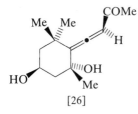

$$MeCH_2{-}CH\overset{Z}{=}CH{-}C{\equiv}C{-}CO{-}\underset{O}{\langle\!\!\!\rangle}{-}CH\overset{E}{=}CH{-}CO_2Me \qquad [22]$$

A number of naturally occurring allenes are also shown, and some of these have biological activity. Probably the best known example is mycomycin [23], a potent antibiotic but toxic, and in common with other polyacetylenes, very unstable to light and in air. The allene [24] has antifungal properties, while compound [25] is a sex attractant of the male Dried Bean Beetle. Finally, the highly functionalized metabolite [25], commonly called 'grasshopper ketone', is utilized by a species of large, flightless grasshoppers, to repel predatory ants. This compound is, however, of isoprenoid origin.

$$HC{\equiv}C{-}C{\equiv}C{-}CH{=}C{=}CH{-}CH\overset{Z}{=}CH{-}CH\overset{E}{=}CHCH_2CO_2H \ [23]$$

$$HOCH_2CH{=}C{=}CH(CH_2)_3CO_2Me \qquad [24]$$

$$Me(CH_2)_6CH_2CH{=}C{=}CHCH\overset{E}{=}CHCO_2Me \qquad [25]$$

[26]

Biogenesis
It was fairly readily established that polyacetylenes are derived from acetyl-SCoA (one mole) and malonyl-SCoA: they were labelled, as

expected, on alternate carbon atoms when [1-^{14}C]-acetate was used as precursor e.g. [27].

$$\text{Me}\overset{*}{\text{C}}\equiv\text{C}-\overset{*}{\text{C}}\equiv\text{C}-\overset{*}{\text{C}}\equiv\text{C}-\overset{*}{\text{C}}\text{H}\overset{E}{=}\text{CH}\overset{*}{\text{C}}\text{O}_2\text{Me} \qquad [27]$$

Many acetylenes retain the Z-double bond of oleic acid, and the terminal octanoate unit, e.g. [28], and it was suggested that oleic acid

$$\text{Me}(\text{CH}_2)_3\text{CH}{=}\text{CH}-\text{C}{\equiv}\text{C}-\text{C}{\equiv}\text{C}-(\text{CH}_2)_7\text{CO}_2\text{H} \qquad [28]$$

was modified by introduction of unsaturation 'to the left of' the central double bond, and by oxidative degradation of its octanoate unit. Two biosynthetic routes have in fact been established, and these are shown in Fig. 2.5. It is interesting that two rather similar metabolites,

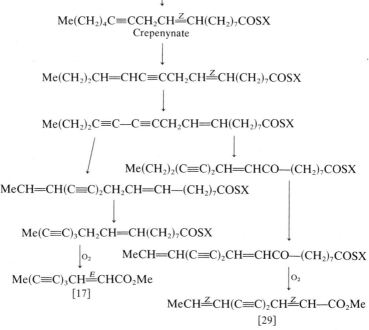

Stearate → Oleate

$$\text{Me}(\text{CH}_2)_4\text{CH}\overset{Z}{=}\text{CHCH}_2\text{CH}\overset{Z}{=}\text{CH}(\text{CH}_2)_7\text{COSX}$$
Linoleate

$$\text{Me}(\text{CH}_2)_4\text{C}{\equiv}\text{CCH}_2\text{CH}\overset{Z}{=}\text{CH}(\text{CH}_2)_7\text{COSX}$$
Crepenynate

$$\text{Me}(\text{CH}_2)_2\text{CH}{=}\text{CHC}{\equiv}\text{CCH}_2\text{CH}\overset{Z}{=}\text{CH}(\text{CH}_2)_7\text{COSX}$$

$$\text{Me}(\text{CH}_2)_2\text{C}{\equiv}\text{C}-\text{C}{\equiv}\text{CCH}_2\text{CH}{=}\text{CH}(\text{CH}_2)_7\text{COSX}$$

$$\text{Me}(\text{CH}_2)_2(\text{C}{\equiv}\text{C})_2\text{CH}{=}\text{CHCO}-(\text{CH}_2)_7\text{COSX}$$

$$\text{MeCH}{=}\text{CH}(\text{C}{\equiv}\text{C})_2\text{CH}_2\text{CH}{=}\text{CH}-(\text{CH}_2)_7\text{COSX}$$

$$\text{Me}(\text{C}{\equiv}\text{C})_3\text{CH}_2\text{CH}{=}\text{CH}(\text{CH}_2)_7\text{COSX}$$

$$\bigg\downarrow \text{O}_2 \qquad \text{MeCH}{=}\text{CH}(\text{C}{\equiv}\text{C})_2\text{CH}{=}\text{CHCO}-(\text{CH}_2)_7\text{COSX}$$

$$\text{Me}(\text{C}{\equiv}\text{C})_3\text{CH}\overset{E}{=}\text{CHCO}_2\text{Me}$$
[17]

$$\bigg\downarrow \text{O}_2$$

$$\text{MeCH}\overset{Z}{=}\text{CH}(\text{C}{\equiv}\text{C})_2\text{CH}\overset{Z}{=}\text{CH}-\text{CO}_2\text{Me}$$
[29]

Fig. 2.5

dehydromatricaria ester [17] and matricaria ester [29] are formed by different routes. Such duplication is a recurrent theme of secondary metabolism.

A possible mode of oxidative degradation is shown below. It involves autoxidation, followed by oxidation of alcohol to ketone, and then β-oxidation (eqn 2.8).

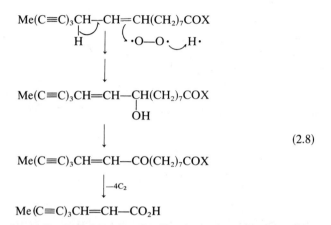

$$(2.8)$$

Modification of the methyl terminus is of less importance, but loss of the terminal carbon atom is quite common. This presumably proceeds via the sequence $-CH_2OH \rightarrow -CHO \rightarrow -CO_2H$ followed by decarboxylation to yield a terminal acetylene.

Most other modifications can be understood in terms of allylic oxidations (like the one depicted in eqn 2.8), or as additions of nucleophiles to the triple bonds. Thus, polyacetylenic thiophenes and enol lactones may be formed as depicted in eqns (2.9) and (2.10), respectively.

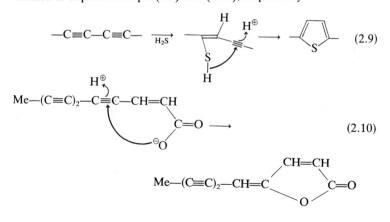

Finally, it should be noted that, as yet, no purified synthetase enzymes have been obtained, and the mode of formation of the triple bond itself is not known. It appears, however, that it is not formed by elimination of phosphate from an enol phospate (eqn 2.11).

$$
\begin{array}{c}
-\text{C}=\text{CH}- \\
\mid \\
\text{O}-\text{P}{=}\text{O} \\
\diagup \quad \diagdown \\
\text{HO} \quad \text{OH}
\end{array}
\ \xrightarrow{\ \times\ }\
-\text{C}\equiv\text{C}- \ +\ \text{HOP(OH)}_2
\qquad (2.11)
$$

'Odd' fatty acids

There are a few naturally occurring fatty acids which possess extraneous methyl groups or cyclopropyl rings. In many instances, these extra methyl groups are present because a starter unit other than acetate was employed, e.g. propionate, isobutyrate, isovalerate, etc. Sometimes, however, it is necessary to postulate that S-adenosyl methionine is the source of the extra methyl group, or methylene ($-\text{CH}_2-$). Thus, Bu'Lock has suggested that tuberculostearic acid [30] is formed from oleic acid and S-adenosyl methionine (eqn 2.12). The cyclopropane acids

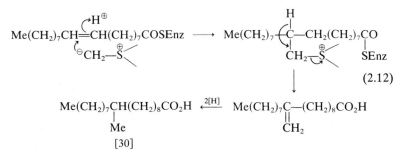

$$(2.12)$$

may then be formed by an analogous process, and this is depicted in eqn (2.13) for lactobacillic acid [31]. Both of these acids are of bacterial origin, [30] from the bacterium responsible for tuberculosis, and [31]

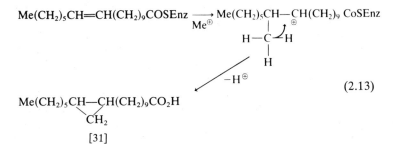

$$(2.13)$$

from a lactobacillus; it is a common finding that bacteria contain 'odd' and often unique secondary metabolites.

·One final example, which demonstrates that a mechanism proposed (or proven) for one metabolite may not hold for the synthesis of a similar compound, is given in eqn (2.14). The C_{19} fatty acid sterculic acid [32] is obtained from nuts produced by a tropical tree. It is not formed by an

$$Me(CH_2)_7C\equiv C(CH_2)_7COSEnz \ \xrightarrow{\times} \ Me(CH_2)_7C\!\!=\!\!C(CH_2)_7CO_2H$$

$$CH_3\!-\!\overset{\oplus}{S}\!\!\diagup \qquad\qquad\qquad [32] \quad CH_2$$

$$Me(CH_2)_7CH\!\!=\!\!CH(CH_2)_7COSEnz \rightarrow Me(CH_2)_7CH\!\!-\!\!CH(CH_2)_7COSEnz$$

$$CH_3\!-\!\overset{\oplus}{S}\!\!\diagup \qquad\qquad\qquad\qquad\qquad CH_2$$

$$\Big\downarrow{}_{-2[H]} \qquad\qquad (2.14)$$

$$[32]$$

extension of the mechanism depicted in eqn (2.13), but rather by oxidation of the cyclopropane fatty acid formed from oleic acid.

Prostanoids and leukotrienes

Further metabolism of eicosatrienoic acid [33] and eicosatetraenoic acid (arachidonic acid) [34] gives rise to prostaglandins, and the major skeletal types are shown in Fig. 2.6. The primary prostaglandins possess a functionalized cyclopentane ring, with seven- and eight-carbon side-chains. In addition, all naturally occurring prostaglandins have a terminal carboxyl or carboxylate group, together with a 13,14-(E)-double bond, and 15-hydroxyl group (usually of S-stereochemistry). These structural elements appear to be essential for maximum biological activity.

The existence of prostaglandins was first noted in 1930, when two New York gynaecologists found that an unidentified substance (or substances) in human semen would cause smooth muscle (i.e the muscle of the intestinal tract, the spleen, uterus, and blood vessels) to contract. Later, Von Euler and Goldblatt confirmed this finding, and demonstrated that these substances would also lower the blood pressure of animals. Von Euler coined the name prostaglandins, since he believed that they derived from the prostate gland. Since then prostaglandins have been shown to be widely distributed in animal tissues, but, with rare exceptions, they are not encountered outside the animal kingdom.

The structures of two prostaglandins were established in 1962 by Bergstrom, using milligram quantities of active substance obtained from several tonnes of sheep seminal vesicular glands. Subsequently, many total syntheses of prostaglandins have been recorded (notably by

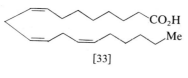

[33]

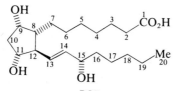

PGF$_{1\alpha}$ PGE$_1$

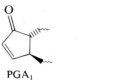

PGA$_1$ PGC$_1$ PGB$_1$

PGD$_1$

[34]

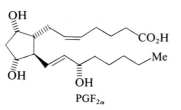

PGF$_{2\alpha}$ PGE$_2$

PGA$_2$ PGC$_2$ PGB$_2$ PGD$_2$

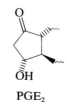

Fig. 2.6

Corey). This has led to a greater availability of these interesting metabolites, and to a consequent explosion in the amount of biological and medicinal research carried out on the compounds. Some idea of the therapeutic potential of these compounds, is given below.

PGE_2 and PGF_{2a} can be used to induce abortion during the first and second trimesters (early and middle stages) of pregnancy. They will also induce labour at full term, and their efficacy in this respect approximates to that of the more commonly used oxytocin (a polypeptide hormone). It is not improbable that prostaglandin analogues could have utility as agents of fertility control.

PGE_1 and PGE_2 are also implicated in the normal functioning of the respiratory system. In particular asthma may be due, at least in part, to aberrant control of the biosynthesis or metabolism of prostaglandins. Both appear to have an essential role in the control of blood pressure, and they may have therapeutic potential as anti-hypertensive agents. They are also effective in decreasing gastric secretion, and may be involved in the control of gastric homeostasis. Indeed, it has been suggested that the damaging effect of aspirin on the stomach may be due to its proven ability to inhibit the biosynthesis of prostaglandins.

In general the PGFs are less potent than the PGEs, except as luteolytic agents, while the most interesting activity of PGD_2 is its ability to inhibit the aggregation of blood platelets. Prostaglandins of the A and C series are not now believed to occur naturally, but are unstable products of further metabolism; and the PGBs have weak activities at best.

In all these actions, the prostaglandins are effective at concentrations of as little as 10^{-6}g/kg of tissue, and generally have a very high rate of turnover, being degraded as follows:

(a) oxidation of the C-15 hydroxyl;
(b) reduction of the 13,14 double bond;
(c) oxidative degradation of the carboxyl and methyl side-chains.

The prostaglandins are produced by most mammalian cells, but have very limited occurrence elsewhere. One intriguing discovery was that a coral of the sea-fan variety (*Plexaura homomalla*) contains $1 \cdot 3$ per cent (dry weight) of epi-PGA_2 (the epimer at C-15, 15-R-hydroxy) and its 15-acetoxy, methyl ester. Methods of converting the 15(R)-compounds into the more active 15(S)-compounds, and thence into PGE_2 and $PGF_{2\alpha}$ have been developed, and this new source of prostaglandins is, at present, being exploited. This venture is now a commercial reality, but on ecological grounds, the project could be disastrous, since excessive or careless harvesting may damage irreparably the ecology of the Caribbean coral reefs. It would be surprising if these prostaglandins were inessential for the coral, but possible functions are as yet merely speculative.

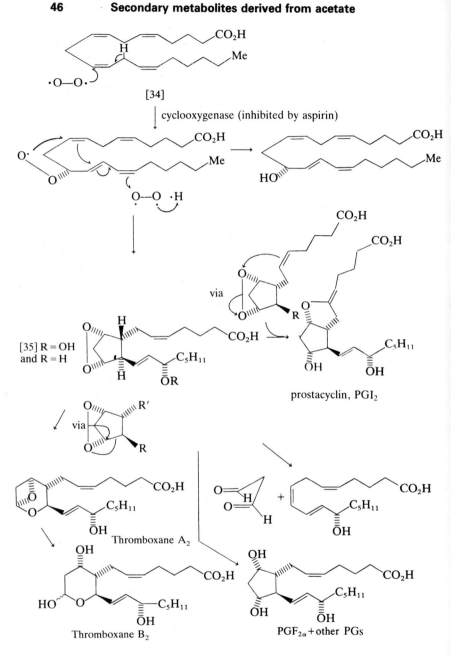

Fig. 2.7

Biogenesis

It is now well established that prostaglandins of the 'one' and 'two' series are derived from eicosatrienoic acid [33] and eicosatetraenoic acid (arachidonic acid) [34], respectively. The labelled acids were incorporated into prostaglandins by extracts from sheep seminal vesicular glands which had prostaglandin synthetase activity. However, the actual mode of biosynthesis is unknown, although several intermediates have been isolated. The endoperoxide [35] (R = OH) is a particularly interesting intermediate, and may be converted into PGE_2 by a homogenate of sheep vesicular glands. A reasonable biosynthetic pathway is shown in Fig. 2.7.

The biosynthetic pathway involves cyclization to produce the endoperoxides [35], and thence the primary prostaglandins. Thromboxane A_2 has a biological half-life of around 30 s, and its structure is based upon products of its metabolism, such as thromboxane B_2. It is a potent bronchoconstrictor and causes aggregation of platelets. By contrast, prostacyclin, PGI_2, with a biological half-life of 5 min, is a potent vasodilator and inhibitor of platelet aggregation. Clearly the release of thromboxane A_2 by platelets, and PGI_2 by blood vessels, help to control platelet aggregation; and there is much interest in analogues of PGI_2 for the control and treatment of arterial disease and prevention of thrombosis.

Other pathways operate with arachidonate as substrate, and the most important of these gives rise to the leukotrienes (Fig. 2.8). These species are highly active (at the nanogram level) mediators of various immune responses, and are intimately involved in allergic states. Antagonists are thus a prime target of pharmaceutical research.

Cyclization of linear polyketides: formation of polyphenols

We shall now consider the large group of secondary metabolites which are formed via cyclization of linear polyketide chains. These compounds contain, almost without exception, aromatic rings, and one or more phenolic hydroxyl groups. Most of these compounds are derived solely from acetate and malonate, but in marked contrast to the biosynthetic pathway to fatty acids and derivatives, there is little or no reduction of the keto functionalities. It has been suggested that the growing β-polyketo thioester chain is stabilized by hydrogen bonding to the synthetase enzyme, or by chelation of its semi-enolate with a metal ion held by the enzyme [36]. It is likely that while so held, some of the keto groups may be reduced (NADPH also held by the enzyme), and alkylation of some of the active methylene groups (S-adenosyl methionine) could also occur. Although it is known that acyl-SCoA species are involved, the exact nature of enzyme-bound intermediates has yet to be established.

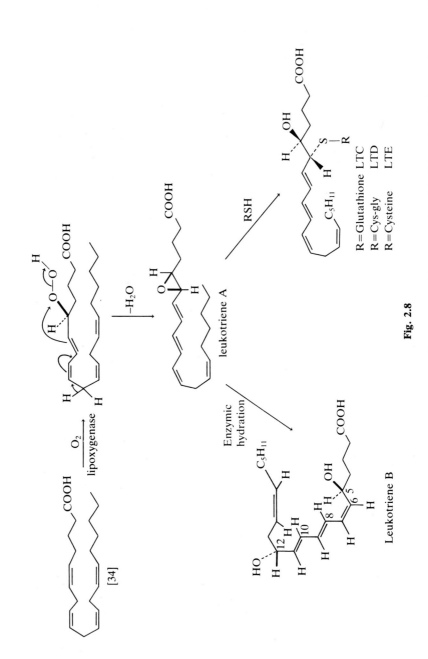

Fig. 2.8

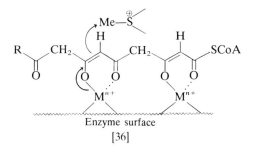

[36]

At some stage, coiling of the polyketide chain leads to intramolecular aldol or Claisen condensations, to produce phenols. Thus the biosynthesis of orsellinic acid [37], a constituent of many fungi and lichens, may proceed as shown in eqn (2.15). The closely related metabolite, phlora-

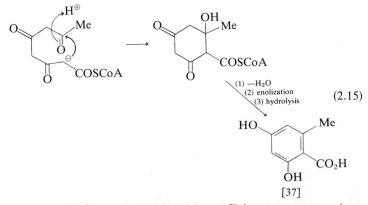

$$(2.15)$$

[37]

cetophenone [38] may be produced by a Claisen-type process (eqn 2.16). Once more we have utilized our knowledge of the quite remarkable reactivity of the thio-CoA functionality, in order to propose reasonable mechanisms. We know also that enzymes utilize (with unerring accuracy)

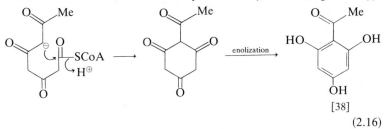

[38]

$$(2.16)$$

the steric and electronic properties of their substrates, and mechanisms are usually written so that they obey the simple stereoelectronic rules of physical organic chemistry. But how was the polyketide pathway established?

Delineation of the polyketide pathway

As early as 1907, Collie demonstrated the *in vitro* cyclization of polyketides, and suggested that a similar process may operate *in vivo*. He treated diacetyl acetone [39] with base, and obtained a naphthalene derivative (eqn 2.17), while cyclization of dehydracetic acid [40] (from pyrolysis of acetoacetic ester) yielded orsellinic acid [37] (eqn 2.18).

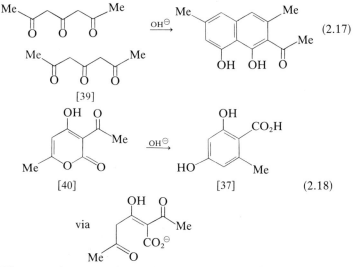

His suggestions were ignored until Birch, in 1953, reexamined the 'acetate hypothesis', and, in a series of elegant experiments, established that a wide range of structural types were derived from acetate (or as we now know, from acetate and malonate). He first of all examined the biogenesis of 6-methylsalicyclic acid [41], which is produced by many microorganisms and by certain higher plants. Birch showed that $[1-{}^{14}C]$-acetate was incorporated, and degradation of [41] gave a labelling pattern consistent with the hypothesis (eqn 2.19).

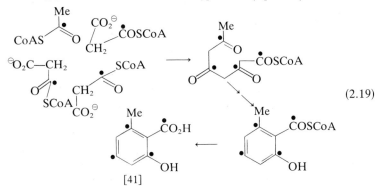

The more complex metabolite, griseofulvin [42], an antifungal and antibiotic substance isolated from *Penicillium griseofulvum*, was also studied by Birch. His ^{14}C-labelling pattern has recently been confirmed by ^{13}C-labelling and ^{13}C-n.m.r. techniques (*vide infra*), and the cyclization must proceed as shown in eqn (2.20). The methyl groups of the methyl ether moieties are derived from S-adenosyl methionine (denoted as C_1).

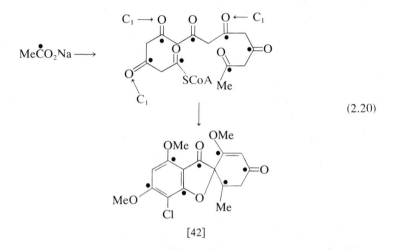

$$(2.20)$$

[42]

That malonate was obligatory was established by feeding [2-^{14}C]-malonate to a *Penicillium* species, and isolating isotopically labelled 6-methylsalicylic acid [41]. As expected, the C_2 unit derived from acetate contained very little activity (eqn 2.21). In contrast, when [2-^{14}C]-acetate was fed, the methyl carbon was also labelled, and almost

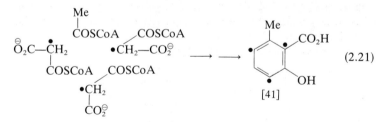

$$(2.21)$$

equal labelling of the four active carbon atoms was recorded. Such asymmetry of labelling can be useful when the position of the starter unit is in doubt, since the starter unit is not labelled when [^{14}C]-malonate is used.

One of the most impressive features of the 'acetate hypothesis' was its predictive capabilities. Many partial structures of metabolites were known, for which ambiguity of total structure existed. Consideration of their probable biogenetic derivation from acetate often led to assignment of an unique total structure. With the advent of ^{13}C-techniques, this approach has become even more useful, and we shall now discuss these, at least in outline.

^{13}C-methods

The isotope ^{13}C has a nuclear spin $I = \frac{1}{2}$, and thus n.m.r. signals may be observed. With the advent of sensitive instruments and the means to accumulate (and thus enhance) weak signals, the amount of data available on ^{13}C-chemical shifts and coupling constants (J_{C-C} and J_{C-H}) is increasing rapidly. Thus, although the natural abundance of ^{13}C is only $1 \cdot 1$ per cent, it is now relatively easy to obtain ^{13}C-spectra for any compound.

One great advantage of ^{13}C-n.m.r. is the considerable spread of chemical shifts. Compare; for example, the ^{1}H- and ^{13}C-spectra for the steroid androst-4-ene-3,17-dione (Spectra $2 \cdot 1$ and $2 \cdot 2$). These shifts are in the same relative order as proton shifts, i.e. alkyl carbons have shifts at higher field (smaller δ value) than aromatic carbons, and aromatic carbons carrying oxygen substituents have larger δ values than those bearing carbon or hydrogen substituents, the carbonyl carbon usually provides the lowest field signal. Assignment of spectra is now almost routine, since tables of typical shifts for ^{13}C-resonances are available.

Feeding experiments using ^{13}C-labelled acetate, propionate, or methionine (up to 99 per cent enrichment in ^{13}C), can lead to isotopically labelled metabolites that give spectra in which certain resonances are enhanced, while others are not. If the spectrum of unlabelled metabolite is available, and has been assigned, the tracer pattern may be discerned at once. However, good incorporations are even more important here than with other isotopes, and most experiments have utilized fungal cultures with which consistently efficient use of tracer is possible.

Another technique that has been used to great effect is the administration of doubly-labelled acetate, i.e. [1,2-^{13}C]-acetate. In this way all carbon atoms of the metabolite are labelled, but since adjacent ^{13}C-nuclei will exhibit C—C coupling, some resonances will be split into doublets. By comparing the observed coupling constants, the adjacent nuclei may be identified, and can be assumed to derive from an intact acetate unit.

Thus, a chain of carbon atoms derived from [1,2-^{13}C]-acetate will show the labelling pattern:

$$\overset{*}{Me}-\overset{\bullet}{C}O_2H \rightarrow \quad \underset{1}{\overset{*}{C}}-\underset{2}{\overset{\bullet}{C}}-\underset{3}{C}-\underset{4}{C}- \quad \text{and} \quad \underset{1}{C}-\underset{2}{C}-\underset{3}{\overset{*}{C}}-\underset{4}{\overset{\bullet}{C}}$$

The signals due to C-1 and C-2 will appear as two pairs of doublets with identical J values (J_{12}). Similarly, C-3 and C-4 will produce doublets with the same J values (J_{34}). However, since it is statistically unlikely that labelled acetate units will be incorporated consecutively into any given

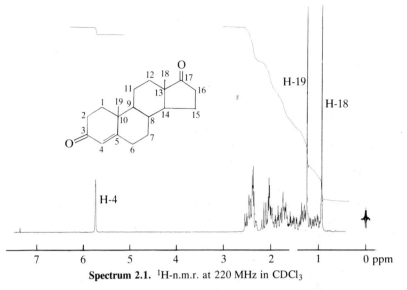

Spectrum 2.1. ^{1}H-n.m.r. at 220 MHz in CDCl$_3$

molecule (there are always many unlabelled acetate molecules in an organism), there will be negligible coupling between C-2 and C-3. Sets of coupled carbon atoms should thus be discernible, and such results provide evidence for the intact incorporation of C$_2$ units. There is usually some observable ^{13}C-H coupling, and this is often of some utility in elucidation of structure, but the spectra are usually proton-decoupled when ^{13}C-^{13}C coupling data are required.

This approach was used to confirm the structure of dihydrolactumcidin, from *Streptomyces reticuli* (see page 56).

(1) The natural abundance ^{13}C-n.m.r. spectrum was obtained, and the probable environment of each carbon atom was predicted from the observed chemical shifts (see Table).

Carbon	Value (ppm)	J value when [1,2-^{13}C]-acetate was used (Hz)	J value when a 1 : 1 mixture of [1-^{13}C]- and [2-^{13}C]-acetates were used (Hz)
2	39.6	35.4	
3	25.9	35.4	43.4
4	59.2	30.7	43.4
4a	64.1	30.7	46.0 and 57
5	140.1	53.4	78.7 and 57.9
6	131.7	53.0	68.1
7	133.8	43.4	68.3
7a	63.1	43.8	46.1
8	114.9	44.8	78.7
9	14.0	44.8	

Peak	1	2	3	4	5	6	7	8	9	10	11	12
ppm	220.07	198.96	170.25	123.98	78.72	77.29	75.86	53.71	50.72	47.38	38.56	35.64
Assignment	C – 17	C – 3	C – 5	C – 4		CDCl₃		C – 9	C – 14	C – 13	C – 10	C – 16 & C – 1

Peak	13	14	15	16	17	18	19	20	21	22
ppm	35.05	33.85	32.48	31.21	30.66	21.66	20.26	17.30	13.66	0
Assignment	C – 8	C – 2	C – 6	C – 7	C – 12	C – 15	C – 11	C – 19	C – 18	

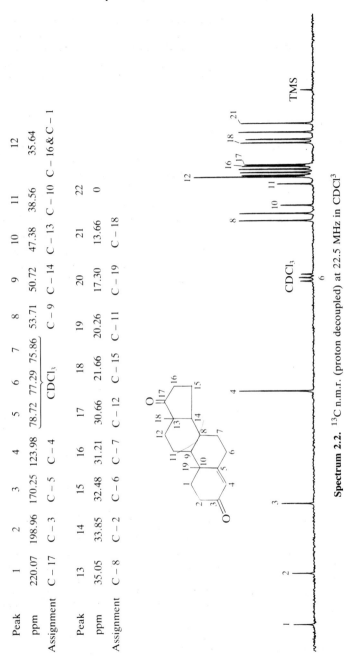

Spectrum 2.2. ¹³C n.m.r. (proton decoupled) at 22.5 MHz in CDCl₃

(2) A polyketide origin was assumed, and [1,2-^{13}C]-acetate, and a 1:1 mixture of [1-^{13}C]-acetate and [2-^{13}C]-acetate were incorporated in separate experiments. The enriched samples of dihydrolactumcidin provided ^{13}C-n.m.r. spectra in which certain of the assigned carbon atoms could be shown to be coupled with other carbon atoms. The possible modes of coupling were:

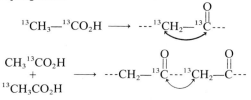

complementary information was thus obtained (see Spectra 2·3 and 2·4).

It can be readily seen that most of the signals appear as triplets in the proton-decoupled spectrum. The centre signal is due to the natural abundance ^{13}C-atoms, while the doublets represent enrichment with ^{13}C and associated coupling to neighbouring enriched centres. Note that in Spectrum 2.4 carbons 4a and 5 couple to two carbons whilst carbons 2 and 9 do not exhibit couplings.

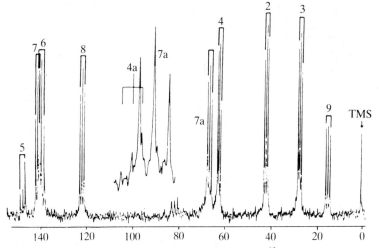

Spectrum 2.3. Proton noise-decoupled Fourier transform ^{13}C-n.m.r. spectrum of dihydrolatumcidin in CDCl$_3$ from ^{13}CH$_3$^{13}CO$_2$Na (90 per cent enriched), 55 mg. The precursor, diluted with unlabelled CH$_3$CO$_2$Na by 2.5 times in order to avoid excess labelling, which would result in a complicated spectrum, was added to the fermentation broth of *S. reticuli* var. *latumcidicus* 14, 19, and 24 h after inoculation. A 100-mg portion of the acetate was added to each 500-ml flask containing 100 ml of medium. The labelled dihydrolatumcidin was isolated by solvent extraction 63 h after inoculation; pulse width, 25 μs; acquisition time, 0.8 s; 11 735 transients. (From *J. Am. Chem. Soc.*, 1973, **95**, 8461–8462.)

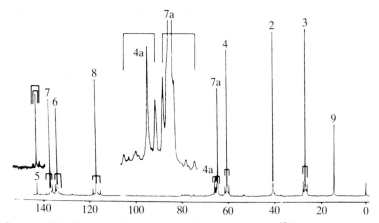

Spectrum 2.4. Proton noise-decoupled Fourier transform ^{13}C-n.m.r. spectrum from a 1:1 mixture of ^{13}CH$_3$CO$_2$Na and CH$_3$^{13}CO$_2$Na (both 90 per cent enriched). A 100-mg portion of the precursor was added to each flask 14 h after inoculation. A yield of 63 mg/five flasks was obtained; 62 325 transients. (From *J. Am. Chem. Soc.*, 1973, **95**, 8461–8462.)

(3) A structure was proposed, and this can be seen to be fully consistent with the coupling data given and with the proposed biosynthetic pathway (eqn 2.22). The enriched sample obtained

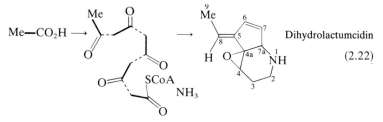

$$(2.22)$$

following incorporation of [1,2-^{13}C]-acetate showed coupling between C-2 and C-3, C-4 and C-4a, C-5 and C-6, C-7 and C-7a, and between C-8 and C-9. This follows the anticipated pattern for incorporation of intact C$_2$ units (see diagram). The enriched sample from incorporation of the mixture of labelled acetates showed coupling between C-3 and C-4, C-5 and C-8, C-4a and C-7a, and between C-6 and C-7. This is the expected complementary pattern, and, in particular, there is no coupling between C-2, C-9, and other carbon atoms: these are terminal carbon atoms in the proposed polyketide, and in consequence coupling is not expected.

Deuterium and tritium n.m.r.
Both of these hydrogen isotopes can be used in n.m.r. experiments, though their full potential has yet to be realized. Deuterium has a nuclear

spin $I = 1$, and a natural abundance of 0.016 per cent; while tritium has a nuclear spin $I = \frac{1}{2}$, and negligible natural abundance. When present in compound, tritium provides sharp signals, with chemical shifts and coupling constants similar to those observed with hydrogen (protium) itself. Deuterium, in contrast, produces broad signals with resonance frequencies, and coupling constants only about one-sixth those of hydrogen. In consequence, its main use to date has been in conjunction with the ^{13}C-isotope. This involves detection of the upfield shift and the coupling caused when it is directly attached to a ^{13}C-nucleus (α shift), or of the shift (negligible coupling) produced when it is one carbon removed from the ^{13}C-nucleus (β shift). Shifts are also observed when the isotopes ^{18}O and ^{15}N are adjacent to a ^{13}C-nucleus, and doubly labelled precursors are now frequently employed. Examples of the uses of these techniques will be given in subsequent sections.

Polyketides derived from acetate

The great variety of structures encountered is a direct result of two metabolic processes: the initial formation of a few, basic skeletal types, and the diverse secondary transformations of these structures. The skeletal type formed will depend upon the number of C_2 units incorporated, the starter unit (if it is not acetate), and the mechanism of cyclization (aldol, Claisen, or other type). Secondary transformations can be of almost any kind (oxidation, reduction, ring cleavage, alkylation, etc.), and may occur before or after cyclization.

It is probably most convenient to classify polyketides according to the number of C_2 units involved in their biogenesis, and this sytem will be adopted here.

Four C_2 units

We have already encountered several examples of polyphenols which incorporate four C_2 units in their structures, e.g. 6-methylsalicylic acid and orsellinic acid; and it is worth noting that enzymic studies with *Penicillium patulum* have indicated a route to 6-methylsalicylic acid which is probably quite close to the true biosynthetic pathway. In the presence of malonyl-SCoA and NADPH, 6-methylsalicylic acid [41], triacetic acid lactone [43], and fatty acids are produced by an enzyme preparation from *P. patulum*. Clearly one mole of NADPH is required per mole of [41] produced, whilst six moles are required for the production of an eight-carbon fatty acid; and it has been suggested that the size of the 'pool' of NADPH regulates the balance between the polyketide and fatty acid pathways.

In addition, Staunton and coworkers have carried out numerous experiments using ^{13}C- and deuterium-labelled acetates, and the results of these experiments can be interpreted in terms of the pathway shown in Fig. 2.9. A reproduction of the n.m.r. spectrum that they obtained after feeding [2-²H₃, 1-¹³C]-labelled acetate to *Penicillium griseofulvum* is also shown in the figure, and provides a beautiful illustration of the utility of the β-shift technique. Two sets of carbon resonances are shown for 6-methylsalicyclic acid, corresponding to unshifted signals (molecules with an adjacent ¹H-isotope), and shifted signals (molecules with an adjacent ²H-isotope). The anticipated labelling pattern is shown, and the amounts of deuterium actually present at each site could be estimated by deuterium n.m.r. Interestingly, the ratio of deuterium at the carbon centres C-7, C-3, and C-5 was 4.9:1.2:1. This was taken as evidence that the C-7 and C-3 carbon centres suffer less loss of deuterium via enolization than at C-5, and, in particular, the result suggests that the reduction and dehydration occur early in the sequence (this too would minimize loss via enolization).

More recently, feeding experiments with [2-²H₁, 1-¹³C]-acetate (monodeuterated acetate) have revealed that the aromatization is probably stereospecific, and hence under enzymatic control. The main

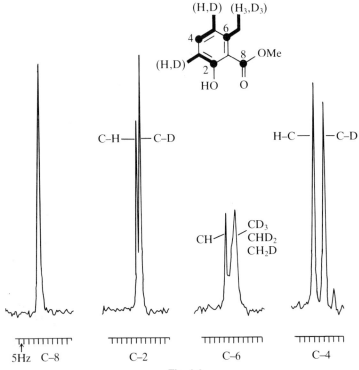

Fig. 2.9

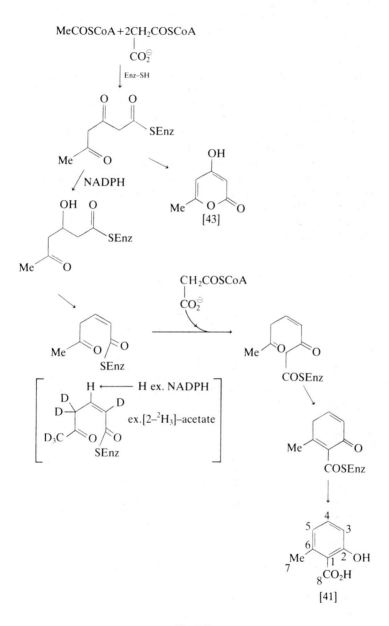

Fig. 2.9

finding was that the relative retention of deuterium at C-3 and C-5 was still 1.2:1, which is inconsistent with a random removal of hydrogen or deuterium via enolization. Isotope effects should ensure that more hydrogen than deuterium would be lost if this was the case, and the ratio would thus change.

Elegant experiments of this kind are beginning to unravel the complexities of polyketide biosynthesis, and the subtleties of the later stages are being revealed.

A rather different mode of cyclization has been observed in the biogenesis of asperlin [44], an antibiotic lactone obtained from *Aspergillus nidulans*. [2-^{13}C]-acetate was incorporated, and the ^{13}C-n.m.r. signal intensities due to carbon atoms 2, 4, 6, 8 and 10 were enhanced relative to those of the remaining atoms. A reasonable pathway is shown in eqn (2.23).

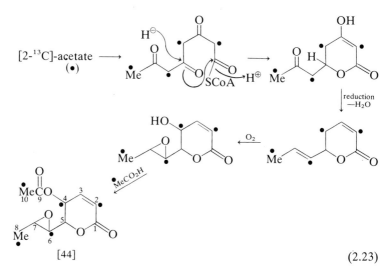

[44]   (2.23)

There are also many examples of polyphenols which contain several aromatic rings, each derived from four C$_2$ units. A particularly rich source of such compounds are the lichens; an alga and fungus growing together in symbiosis. Lichens have been studied rather intensively of late because it is believed that they may be useful as sensitive indicators of levels of environmental pollution by trace metals. They have no root system and derive all nutrients from the atmosphere, hence their potential usefulness. Usnic acid [45] is a rather common lichen metabolite which inhibits tumour growth, and its probable mode of biosynthesis is shown in Fig. 2.10. The two molecules of methylphloracetophenone [46] are joined by a process known as oxidative phenol coupling. This process may be observed *in vitro*, and the structurally similar compound called

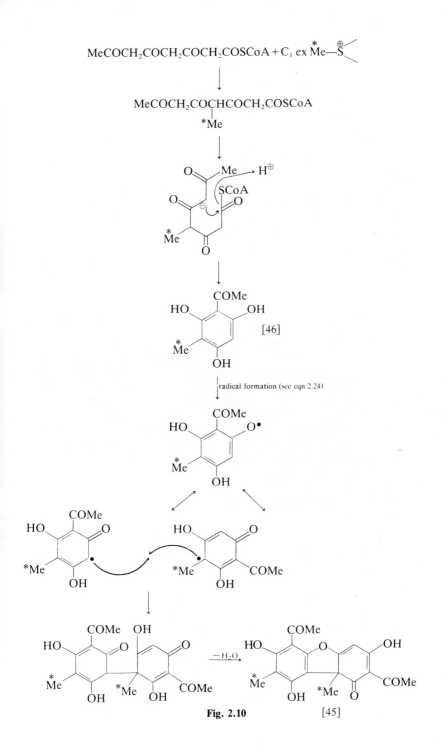

Fig. 2.10

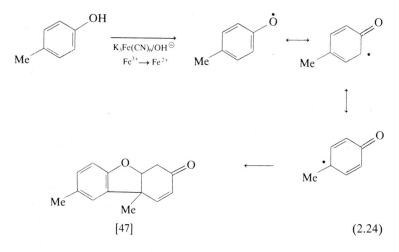

$$[47] \qquad\qquad\qquad (2.24)$$

Pummerer's ketone [47] is produced in this way by treatment of *p*-cresol with potassium ferricyanide (eqn 2.24).

The enzymes responsible for most *in vivo* phenolic coupling are peroxidases. They have a porphyrin as cofactor [48] (i.e. a cyclic, tetrapyrrole, alkylated at the 3 and 4 positions of each pyrrole ring), which is a tetradentate ligand, and uses these four coordination sites (each nitrogen atom) to bind a ferric ion (Fe^{3+}) at the centre of the ring. This ion may accommodate six ligands, and the fifth and sixth coordination positions are occupied by an amino acid residue of the enzyme, and the hydroperoxy ion ($HOO^{\ominus}$) or superoxide ion ($O_2^{\ominus}$), respectively. This structure is not at all unlike that of potassium ferricyanide, and it would be surprising if the enzyme-catalysed process differed markedly in mechanism from that observed *in vitro* with ferricyanide. Other examples of phenolic coupling will be encountered in later sections.

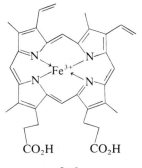

[48]

This kind of secondary transformation is fairly straightforward, and it is usually rather obvious as to which precursor species were utilized. However, many fungal metabolites bear little structural relationship to their progenitors. This is often due to oxidative cleavage of an aromatic ring, and subsequent rearrangement. We shall consider briefly how this cleavage is accomplished.

Oxidative cleavage of aromatic rings

The enzymes responsible for this process are called dioxygenases, and, as the name suggests, they catalyse the incorporation of both oxygen atoms from the oxygen molecule, with concomitant ring cleavage. They almost inviariably require iron (usually Fe^{3+}) as cofactor. Two representative models of cleavage are shown in eqn (2.25). [N.b.: the enzymes known as monooxygenases (mixed function oxidases) which catalyse the incorporation of one atom of a molecule of oxygen in such processes as epoxidation, hydroxylation, and oxidation of nitrogen and sulphur atoms, and which require an electron donor as cofactor, usually NADPH or $FADH_2$, were mentioned in Chapter 1.] Recently, biomimetic oxidative

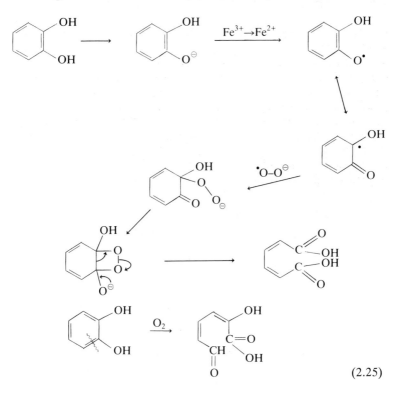

$$(2.25)$$

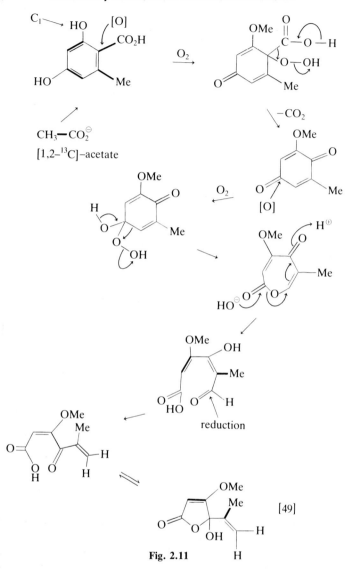

Fig. 2.11 [49]

cleavage of catechol has been carried out using oxygen and cuprous chloride in pyridine containing some methanol, and also using peracetic acid and Fe^{3+}. Both methods gave good yields of products.

Two good examples of oxidative ring cleavage during the biosynthesis of polyketide metabolites are shown in Fig. 2.11. Penicillic acid [49] is an antibiotic and potent carcinogen produced by a variety of fungi, and it is

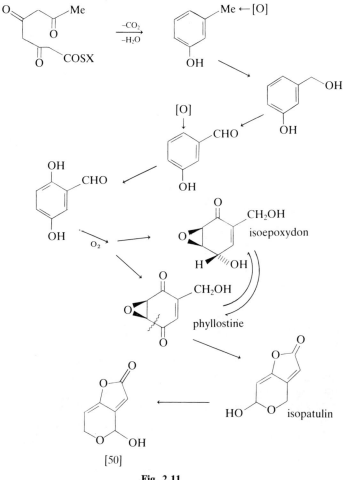

Fig. 2.11

not immediately apparent that it is of polyketide origin. However, isotopic evidence strongly supports the proposed pathway especially the presence of two intact ^{13}C–^{13}C-bonds in the enriched specimen of penicillic acid. Patulin [50] is also a potent carcinogen, and since it occurs in mouldy apples, and is stable in apple products such as pies and juice, it is potentially dangerous to human health. The biosynthesis of patulin has been extensively studied using cell-free preparations from *Penicillium urticae*. The main pathway is shown in Fig. 2.11, but other minor pathways also operate, depending on the nutritional status of the organism. These are just two compounds from amongst several known

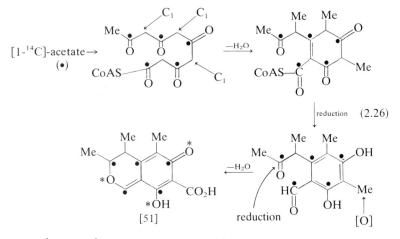

$$[1\text{-}^{14}C]\text{-acetate} \rightarrow$$
$$(\bullet)$$

reduction   (2.26)

[51]   reduction   [O]

carcinogens of common occurrence, which are being studied to determine their possible long-term, deleterious effects on human and animal life. Two other compounds will be mentioned shortly: citrinin and aflatoxin.

Five C_2 units

The biosynthesis of citrinin [51], a carcinogenic antimicrobial metabolite produced by several *Aspergillus* species, and also by *Penicillium citrinum*, has been studied extensively. This very toxic compound was at least partially responsible for the fatalities caused by 'yellow rice disease', an epidemic which occurred in Japan in the early part of this century. The observed labelling pattern was that anticipated for a process involving cyclization of a pentaketide precursor (eqn 2.26) and this has been confirmed using $[1,2\text{-}^{13}C]$-acetate, $[1\text{-}^{13}C, {}^{18}O_2]$-acetate (note *), and other labelled species. The likely pathway to eugenone [52] is more conventional (eqn 2.27).

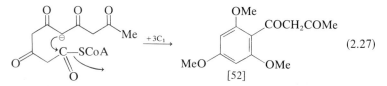

(2.27)

Oxidative secondary transformations are also common, and the fungal tropolones are produced in this way. Thus, sepedonin [53] has been labelled with ^{13}C following incorporation of $[1\text{-}^{13}C]$-acetate and $[2\text{-}^{13}C]$-acetate (separate experiments), and of $[^{13}C]$-methanoate (source of C_1). The overall pattern obtained was consistent with the pathway shown in Fig. 2.12.

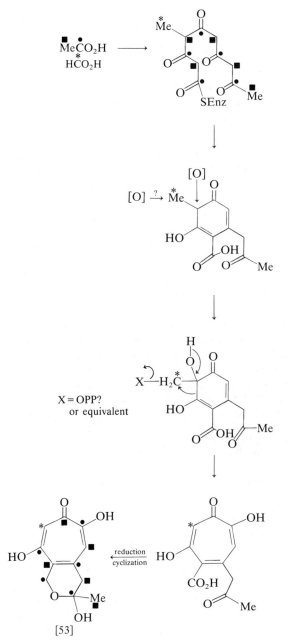

[53]

Fig. 2.12

The structurally similar tropolone, stipitatonic acid [54], is in fact produced from a tetraketide precursor, and a possible pathway is shown in eqn (2.28).

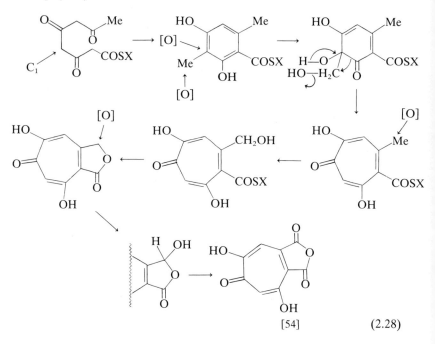

[54]   (2.28)

Six C_2 units

Polyphenols derived from a hexaketide progenitor are somewhat rare. Although some naphthoquinones are derived from acetate, most are formed from shikimate, and via other, more obscure pathways. The compounds plumbagin [55] and 7-methyljuglone [56] (R = Me), however, appeared to be derived from acetate and malonate and possible pathways are shown in (eqn 2.29); but juglone (R = H) derives at least part of its skeleton from shikimate. These are metabolites of certain carnivorous plants, which live on nitrogen-poor soils and depend upon insects as a source of nitrogen. It is suggested that these compounds act as antimicrobial agents, thus preventing the growth of bacteria which might otherwise 'consume' the captured prey before the plant has time to digest it.

It is also interesting to note that juglone [56] (R = H) is largely responsible for the lack of plant growth around walnut trees. It occurs in the leaves, fruit, and other tissues of the tree in the form of hydroxyjuglone (reduced form), which is non-toxic. This is leached from

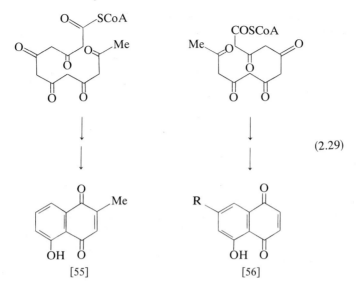

(2.29)

[55]   [56]

the leaves, etc. by rain, and is converted in the soil into juglone, which acts as a growth inhibitor for most other plant species. Such inhibition mediated by chemical species released by plants is known as *allelopathy*, and we shall consider other examples of this phenomenon in due course (primarily in Chapter 7).

The somewhat obscure antifungal agent, variotin [57], is also derived from a hexaketide precursor. ^{13}C-techniques have been used to establish its polyketide origin: the pyrrolidone ring is not derived from acetate (eqn 2.30).

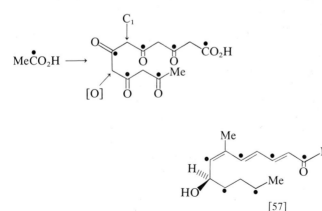

(2.30)

[57]

Seven and eight C₂ units

Many polyphenols are derived from heptaketide and octaketide progenitors. Griseofulvin [42], as we have already seen, is derived from an heptaketide; and many anthraquinones are formed from an octaketide despite the deceptive appearance of oxygen atoms at unexpected positions. An extensive labelling study by Staunton and Leeper has demonstrated that rubrofusarin, [58] (R = H) from *Fusarium culmorum*, is formed from a heptaketide which adopts the folding pattern shown in eqn (2.31) rather than the alternative one. Thus, for example, C-11 and

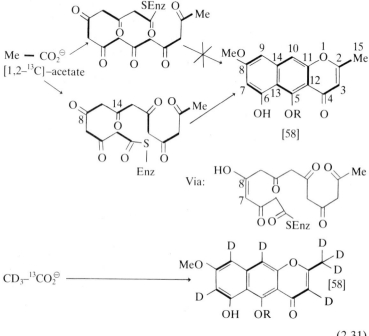

(2.31)

C-12 were mutually coupled in the ¹³C-n.m.r. spectrum, and there was no coupling between C-11 and C-10 or between C-12 and C-5. When [1-¹³C, 2-²H₃]-acetate (i.e. $CD_3^{13}CO_2^-$) was employed, β-shifts were observed for the ¹³C-signals of C-2, C-6, C-8, and C-14, but not for C-11. In the alternative pathway, C-10 and C-11 would be derived from an intact two-carbon unit, and C-11 would be adjacent to a carbon bearing one deuterium atom, which should thus produce a β-shift. (All of these n.m.r. studies were carried out with rubrofusarin methyl ether, R = Me.)

Another interesting facet of this work was the results obtained with deuterated acetate (i.e. $CD_3CO_2^-$). When using deuterium (or tritium)

isotopes there is always the chance that isotope will be lost via keto–enol tautomerism, and in this system there was a greater than expected loss for C-7. The enol form shown in eqn (2.31) would help to direct ring closure between C-13 and C-14, and the extra loss of isotope would be explained by reversion to the keto-form and then regeneration of this key enol.

There was also an unexpectedly high retention of deuterium at C-3, and this suggests that the pyrone ring is formed very early in the biosynthesis (perhaps even before the C-13 to C-14 ring closure). If this was not the case, extensive loss of deuterium via keto–enol tautomerism would be anticipated. This is a further example of how experiments are beginning to shed light on the subtleties of the later stages of polyketide biosynthesis.

The anthraquinone emodin [59] has a regular biogenesis, and an elegant biomimetic synthesis has been reported (eqn 2.32). Alizarin [60], the orange-red pigment isolated from the madder plant, and used as a dye in Persian rugs (before the advent of synthetic dyes), is however derived from shikimate and mevalonate. Its biosynthesis will be discussed in Chapter 6.

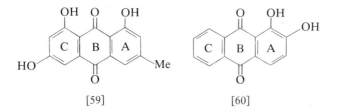

[59]   [60]

It appears that, at least in higher plants, those anthraquinones with substituents on both A and C rings, are derived via the acetate–malonate pathway. Those with only one ring substituted, such as alizarin, derive from shikimate and mevalonate. However, the situation in lower organisms is not so clear-cut, since pachybasin [61], which ought to be derived from shikimate, does in fact arise via the acetate–malonate

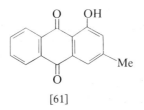

[61]

pathway. This illustrates how careful one must be when making general-izations, especially where widely different organisms are concerned. Nature often seems capricious in her choice of pathways.

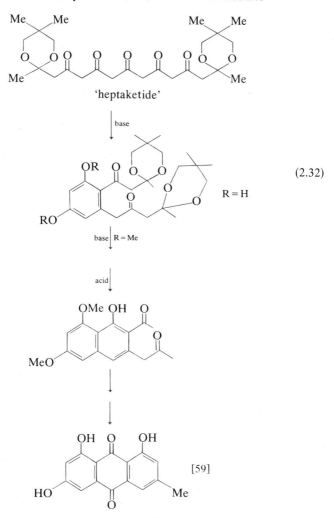

'heptaketide'

(2.32)

R = H

[59]

The anthraquinone ring system appears to be metabolically active, and it has been suggested that anthraquinones may have a role in the process of fruit development in plants. Oxidative ring fission is a common process in fungi. Thus questin [62], is converted into sulochrin [63] (eqn 2.33), and this process is also thought to be involved in the biosynthesis of several xanthones.

The biogenesis of xanthones had been little studied until quite recently. The co-occurrence of some xanthones with anthraquinones or anthrones (one less carbonyl function) suggested that they might be produced by

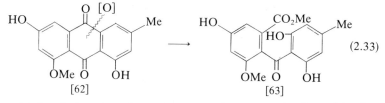

$$(2.33)$$

oxidative fission, broadly analogous to the process implicated in the biosynthesis of sulochrin, especially since a family of fungal pigments, the secalonic acids, e.g. [64], still retain the carboxylate functionality.

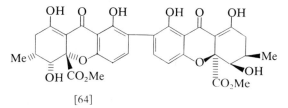

These complex mycotoxins are produced by the fungus *Claviceps purpurea*, and are as much responsible for the symptoms of ergotism as are the well-known ergot alkaloids (see Chapter 6), though this was not realized until the 1960s. Holker, Simpson, and coworkers have shown that an oxidative cleavage occurs in the biosynthesis of tajixanthone (see Fig. 2.13) and related metabolites. ^{13}C-enrichment data were in accord with the pathway shown in Fig. 2.13; and a labelling study using [1-^{13}C]-acetate in conjuction with $^{18}O_2$ was also in accord with this pathway. In this experiment, the ^{13}C-signals were shifted by the presence of adjacent ^{18}C isotopes, and it was thus possible to show that either of two phenolic hydroxyls could be involved in formation of the xanthone ring.

Similarly, the fungal metabolite aflatoxin B_1 [65] is derived from a co-occurring xanthone, sterigmatocystin [66], which in turn is believed to be produced by oxidative fission of the anthraquinone versicolorin A [67] (Fig. 2.14). Aflatoxin B_1 is just one member of a family of metabolites, all of which contain fused bis-dihydrofuran moieties. They are amongst the most toxic compounds known, and were first discovered following the deaths of scores of turkeys, which had fed on mouldy peanuts. Various strains of *Aspergillus* produce aflatoxins, and they are potent carcinogens, causing lesions in the mammalian liver. Studies with primates have shown that the long-term chronic effects are very considerable, and it has been suggested that the high incidence of hepatocarcinoma (liver cancer) in parts of Africa may be due, at least in part, to the utilization of contaminated peanut meal as the staple part of the diet. For example, in the highlands of Kenya where the average intake of aflatoxins is 3.5 ng/kg

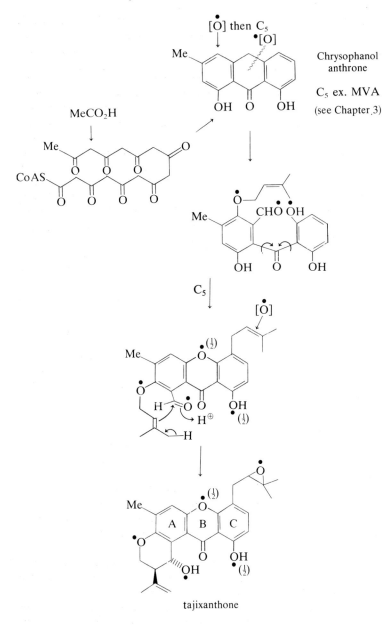

Fig. 2.13

body weight per day, the incidence of liver cancer is 0.7/100 000 people per year. In parts of Mozambique, the corresponding figures are 222 ng/kg body weight per day, and 13 cases/100 000 people per year.

The scheme shown in Fig. 2.14 is in accord with the [^{14}C] (and [^{13}C]) studies of Buchi, Holker, and Tanabe, and with the results obtained by Steyn using [1,2-^{13}C]-acetate: ^{13}C–^{13}C coupling is observed between those carbons at either ends of the bonds shown in bold type. It is envisaged that the anthraquinone [67] is cleaved to produce the xanthone sterigmatocystin [66], which itself suffers oxidative ring fission to yield aflatoxin B$_1$ [65]. Certainly sterigmatocystin can be converted into aflatoxin B$_1$ by cultures of *Aspergillus parasiticus*, but the earlier steps of the sequence must be regarded as hypothetical. The involvement of a decaketide precursor seems most likely, and experiments appear to rule out the alternative mode whereby a heptaketide progenitor incorporates a C$_4$ unit to form the bis-dihydrofuran system (C$_4$ units are poorly incorporated).

On the basis of studies with mutants of *Aspergillus parasiticus*, averufin has been identified as an early intermediate. Townsend and coworkers have prepared both deuterium and ^{13}C-labelled averufin, and have demonstrated their incorporation into aflatoxin B$_1$, but of more interest mechanistically was the observation that a sample of averufin with a deuterated methyl group (CD$_3$-) was converted into versiconal acetate with retention of label. A Baeyer-Villiger-like oxidation has, in conseqence, been proposed for this interconversion. This is shown in Fig. 2.14. Finally, the mechanism must also explain why the bond from the C-1' oxygen to C-5' carbon of averufin remains intact through to versiconal acetate—this was established using [1'-^{18}O, 5'-^{13}C]—averufin.

Nine C$_2$ units

The tetracyclines are probably the best known examples of a class of metabolites derived from cyclization of a nonaketide. Representative examples of medicinally useful tetracyclines are terramycin [68] aureomycin [69], and tetracycline[70]. All are broad spectrum antibiotics, effective against a wide range of organisms, and all can be used internally or topically (as in creams and ointments). Although the mode of biosynthesis is not completely established, it is probable that it proceeds as shown in Fig. 2.15.

This scheme was proposed largely as a result of work carried out with mutant strains of streptomyces. Mutants are obtained when a parent strain is irradiated with X-rays, γ-rays, or treated with chemical mutagens, such as alkylating agents (e.g. dimethyl sulphate). These various treatments lead to random modification of the genetic material (DNA), and usually to the death of the organism. However, often one or more of these genetic modifications will lead to a viable mutant, albeit one which produces

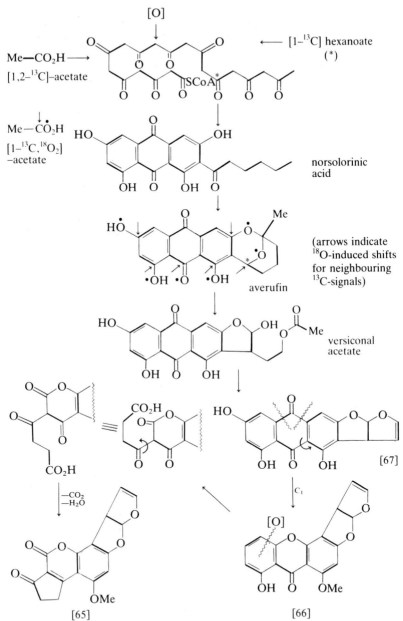

Fig. 2.14

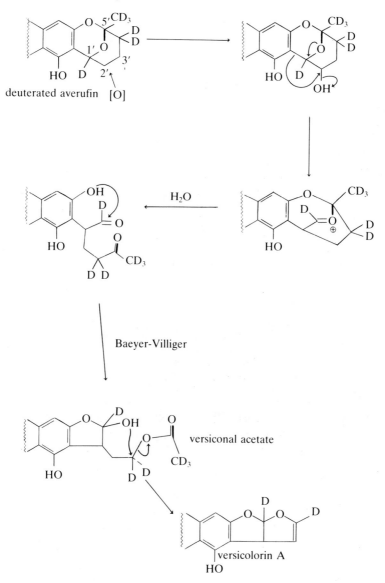

deuterated averufin [O]

H₂O

Baeyer-Villiger

versiconal acetate

versicolorin A

Fig. 2.14

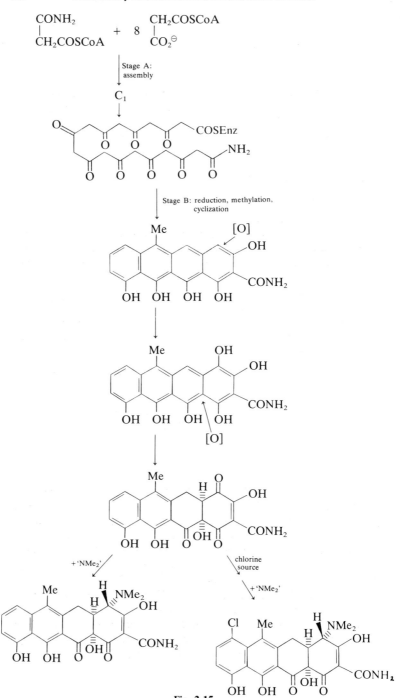

Fig. 2.15

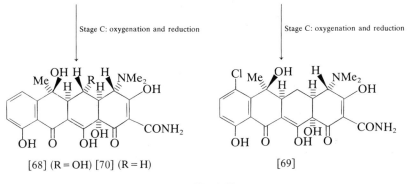

Fig. 2.15

aberrant tetracyclines, or in which normal biosynthesis is impaired. Identification of these mutant strains, and examination of the various impairments, usually leads to deductions concerning the normal biosynthetic pathway.

For example, a mutant in which the malonamido-SCoA starter (probably unique to tetracycline biosynthesis) could not be synthesized (or utilized), led to the utilization of acetoacetyl-SCoA [5] instead, and to production of a whole class of aberrant tetracyclines, i.e. the 2-acetyl series, e.g. [71].

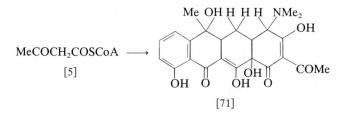

Similarly, 6-nortetracyclines, e.g. [72] (R = H), are produced by mutants lacking the capability for alkylation, that is, introduction of a methyl group from S-adenosyl methionine does not occur. There is thus a defect at state B (Fig. 2.15). Both this mutation and the previous one affect early stages of the pathway, and subsequent steps are not affected. The consequence of a mutation which affects stage C is production of tetracyclines with the incorrect level of oxidation (eqn 2.34).

Finally, tricyclic tetracyclines, e.g. [73], are produced by certain mutant strains, and here the mutation must affect the cyclization step. What this study tells us is that alkylation and reduction must occur before cyclization, since these structural features are present. Not surprisingly, such gross structural aberration is incompatible with further

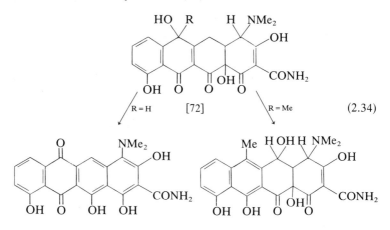

(2.34)

transformation, and this compound is an end-project: it cannot be utilized by the normal enzymes, even if they are present.

The biosynthetic pathway has also been investigated using ^{13}C-labelled acetates, [1, 2, 3-^{13}C]-malonate, and [1-^{13}C, ^{18}O$_2$]-acetate; and there is no doubt that the direction of folding is as shown in Fig. 2.15.

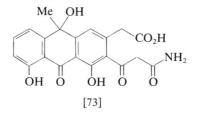

[73]

The anthracyclinones, e.g. daunomycin [74], are also produced by cultures of Streptomyces species, and are of particular interest because of their very potent and broad spectrum anti-cancer activity. Several members of the family and certain synthetic analogues are used extensively in cancer chemotherapy regimes. The biosynthesis of daunomycin proceeds

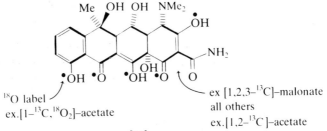

^{18}O label
ex.[1–^{13}C,^{18}O$_2$]–acetate

ex [1,2,3–^{13}C]–malonate
all others
ex.[1,2–^{13}C]–acetate

[68]

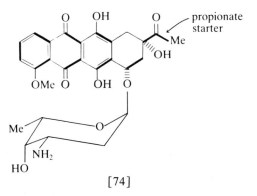

[74]

from one propionate unit and nine acetates and the final acetate unit suffers decarboxylation.

Polyketides which incorporate novel starter units

A large number of polyphenols are derived from malonate units and a starter unit other than acetate. The tetracyclines are one such example: here the starter unit is malonamido-SCoA. Often all members of a particular plant family produce polyphenols which possess this type of structural modification and this has obvious utility as an aid to phytotaxonomic classification. Thus, plants of the family Anacardiaceae use a series of unsaturated fatty acids as starter units, and add three or four units of malonyl-SCoA, to produce, after cyclization, alkenyl polyphenols of the type shown in Fig. 2.16. The plants always contain a mixture of phenols in which the side-chain R is mono-, di-, or triunsaturated, reflecting the fact that a number of unsaturated acyl-SCoA starters can be employed. The family of phenols known as the urushiols are the major irritant constituents of poison ivy: the pure compounds are very active allergens (cause allergic reactions) in Man.

Many lichen metabolites also incorporate starter units other than acetate. Thus, olivetoric acid [75] is probably derived from a C-alkyl- and C-alkanoyl orsellinic acid, as shown in Fig. 2.17. It is interesting to note that most of the complex secondary metabolism of lichens is due to the fungal partner. In the few instances where fungus and alga have been separated and studied in isolation, the fungus alone produces the typical aromatic acids, but no esters of the type shown in Fig. 2.17. It seems that the algal partner is required for coupling to occur. However, it remains to be established whether this secondary metabolism benefits either organism.

Finally, it should be noted that it is sometimes difficult to reconcile labelling patterns with a cyclization involving only one polyketide chain.

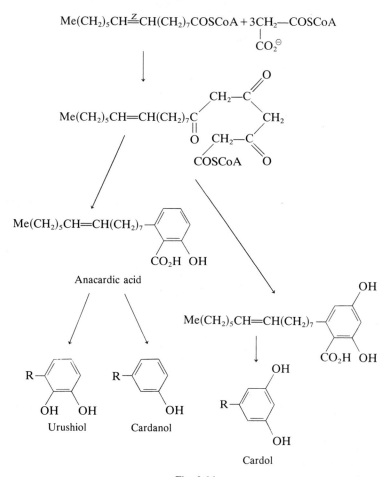

$$Me(CH_2)_5CH\overset{Z}{=\!\!=}CH(CH_2)_7COSCoA + 3CH_2\!-\!COSCoA$$
$$\underset{CO_2^{\ominus}}{|}$$

Anacardic acid

Urushiol

Cardanol

Cardol

Fig. 2.16

The fungal metabolite sclerin [76] and its cometabolite sclerotin A [77] were believed to arise via different pathways (eqn 2.35). However, the results of experiments with [1, 2-^{13}C]-acetate are also consistent with the pathway shown in Fig. 2.18. This was suggested by Staunton, who reasoned that the structural similarity between sclerotin A [77] and citrinin [51] could indicate a parallel biogenesis. The pathway includes a hydrolytic cleavage of [77] (after keto-enol tautomerism), rotation about the central carbon–carbon bond, and finally ring closure. Incorporation of [^{2}H$_3$]-acetate showed that only the methyl group marked with an asterisk contained deuterium, and this is also consistent with the proposed

MeCH$_2$CH$_2$CH$_2$CH$_2$COSCoA + 4CH$_2$—COSCoA
$\qquad\qquad\qquad\qquad\qquad\qquad$|
$\qquad\qquad\qquad\qquad\qquad\qquad$CO$_2^{\ominus}$

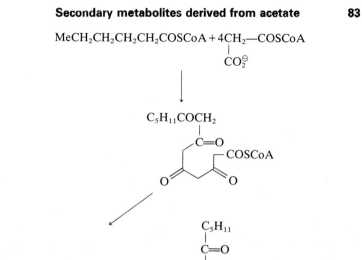

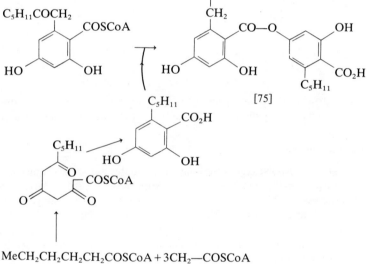

[75]

MeCH$_2$CH$_2$CH$_2$CH$_2$COSCoA + 3CH$_2$—COSCoA
$\qquad\qquad\qquad\qquad\qquad\qquad$|
$\qquad\qquad\qquad\qquad\qquad\qquad$CO$_2^{\ominus}$

Fig. 2.17

route—though it would also be compatible with the original suggestion (eqn 2.35). Finally a sample of ^{14}C-labelled sclerotin A was isolated, and this was shown to act as an advanced intermediate when fed to the microorganism, *Sclerotinia sclerotiorum*, and was converted into sclerin. This provides rather more cogent evidence for the new route. Sclerin itself is of some considerable interest since it functions as a potent

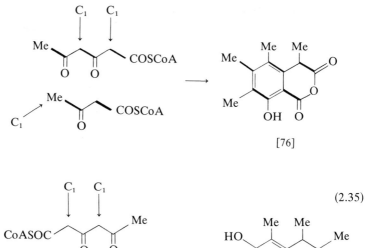

[76]

(2.35)

[77]

plant growth-promoting agent, and the pathogenic fungus presumably exerts its activity by disrupting the normal growth of the infected plant.

Macrocyclic antibiotics

A multitude of polyoxygenated fungal metabolites are known, which are assembled from C_3 units: that is, from propionyl-SCoA [78] as starter and with successive additions of methylmalonyl-SCoA [79]. (Propionyl-SCoA is derived, at least in part, from the degradation of fatty acids of odd-number chain length.) Most of these metabolites possess large rings, and are potent antibiotics: they are commonly known collectively as *macrocyclic* (or *macrolide*) *antibiotics*.

One of the first compounds to be investigated was erythromycin A [80], from *Streptomyces erythreus*. This fungus incorporated [1-], [2-], and [3-^{14}C]-propionate into erythromycin, and extensive degradation produced a labelling pattern which was consistent with the biosynthetic route shown in eqn (2.36). Incorporation of 2-methyl malonate labelled on the methyl carbon, proved conclusively that the methyl groups of erythromycin were derived from this precursor, and not from S-adenosyl methionine; in addition, when [1-^{13}C,^{18}O$_2$]-propionate was employed, ^{13}C signals for those carbons adjacent to an ^{18}O-isoptope were shifted, as anticipated. (See Spectrum 2.5; from *Tetrahedron*, 1983, **39**, 3449–3455.) All of the

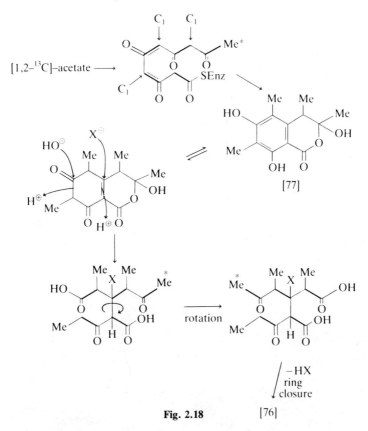

Fig. 2.18   [76]

genes for erythromycin biosynthesis have been cloned—the first time such a feat has been accomplished.

The *ansamycin antibiotics*, of which rifamycin S [81] is a typical example, are also derived, in part, from propionate. The remaining portion of the ring system is known to be derived originally from glucose, via 3-amino, 5-hydroxybenzoic acid, but the exact pathway is unknown. Biosynthesis proceeds in a clockwise direction, as depicted in eqn (2.37). These metabolites have interesting antiviral and antibiotic properties, and also function as inhibitors of RNA tumour virus reverse transcriptases (enzymes responsible for the incorporation of viral genetic information into the host's chromosomes). Although many of them are too toxic for clinical use, they are much used in molecular biology as a research tool.

Butyryl-SCoA units are somewhat rare in Nature, but do occur, along with propionyl-, malonyl-, and acetyl-SCoA units, in metabolite [82] called lasalocid A. Incorporation of [1-^{13}C]-butyrate and [1-^{13}C]-propiionate led to enhanced signals (those anticipated) in the ^{13}C-n.m.r. spectrum of the metabolite, and there is little doubt that the compound is assembled as shown in eqn (2.38).

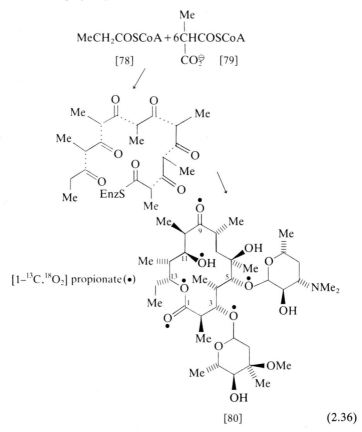

[80]   (2.36)

Observation of isotopically shifted ^{13}C-resonances was the key to elucidating the biosynthetic pathway to the avermectins, e.g. [83] when $[1-^{18}O_2, 1-^{13}C]$-acetate and $[1-^{18}O_2, 1-^{13}C]$-propionate were fed to *Streptomyces avermitilis*. These metabolites are currently of immense interest because they possess potent antiparasitic activity.

The early promise of the Collie–Birch acetate hypothesis has been realized; and with the advent of sophisticated n.m.r. techniques, we have the means by which prediction may become fact with remarkable facility. One final example is particularly instructive since it provides a demonstration of how the technique can be used to establish that oxidative ring cleavage occurs at an intermediate stage. This is the work of Holker,

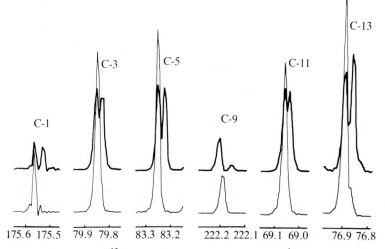

175.6 175.5   79.9 79.8   83.3 83.2   222.2 222.1 69.1 69.0   76.9 76.8

Spectrum 2.5. Partial ^{13}C-n.m.r. spectra of erythromycin A 2'-benzoate showing the signals of the oxygen-bearing carbon atoms of the lactone ring. Lower trace: sample derived from $[1\text{-}^{13}C]$-propionate. Upper trace: sample derived from $[1\text{-}^{18}O_2, 1\text{-}^{13}C]$-propionate.

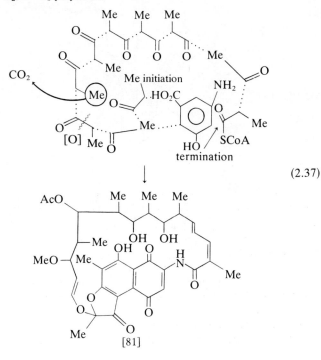

(2.37)

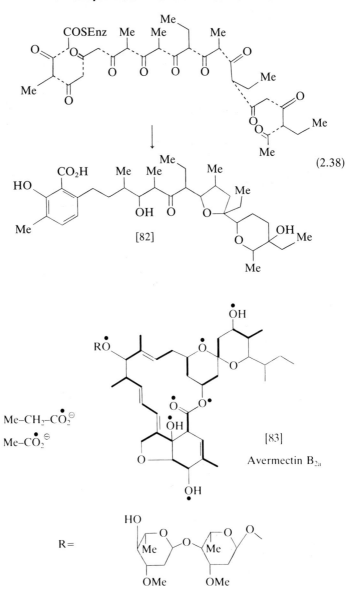

(2.38)

[82]

[83]

Avermectin B$_{2a}$

$Me-CH_2-\overset{\bullet}{C}O_2^{\ominus}$

$Me-\overset{\bullet}{C}O_2^{\ominus}$

R=

Gudgeon, and Simpson on the tetronic acids [84], which are produced by *Penicillium multicolor*. [1-^{13}C]-acetate and [2-^{13}C]-acetate were incorporated, and the n.m.r. spectra were assigned. Finally, those carbon atoms which derive from intact incorporation of acetate units, were identified by labelling with [1,2-^{13}C]-acetate. The overall labelling pattern was:

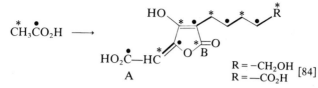

$$CH_3\overset{\bullet}{CO_2H} \longrightarrow$$

R = —CH₂OH [84]
R = —CO₂H

The alternating pattern suggests that the carboxylic acid carbon (A) may have been bonded to the lactone carbon atom (B) at an intermediate stage: there is no C—C coupling between these carbon atoms. A possible intermediate might be [X]:

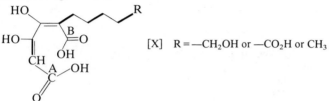

[X] R = —CH₂OH or —CO₂H or CH₃

and known modes of oxidative ring-opening of aromatic compounds

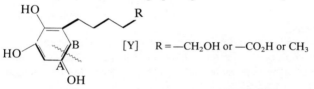

[Y] R = —CH₂OH or —CO₂H or CH₃

suggest precursor [Y]. This could, in turn, be derived from the polyketide precursor [Z] with concomitant loss of CO_2. In addition, ^{14}C-tracer results and an ^{18}O-acetate study make this route even more likely:

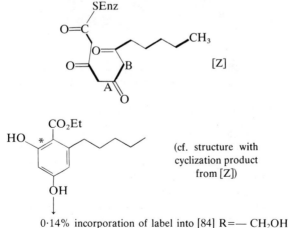

[Z]

(cf. structure with cyclization product from [Z])

0·14% incorporation of label into [84] R=— CH₂OH
0·9% incorporation of label into [84] R=— CO₂H

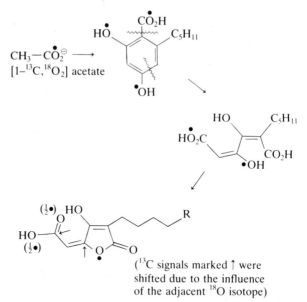

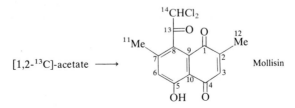

(^{13}C signals marked ↑ were
shifted due to the influence
of the adjacent ^{18}O isotope)

Problems

2.1. There are several possible modes of folding of polyketide precursors, which could yield the metabolite mollisin (ex. *Mollisia caesia*). Write down as many as possible, and then suggest the most likely in light of the following ^{13}C-data:

[1,2-^{13}C]-acetate ⟶

Mollisin

^{13}C-^{13}C coupling observed between C-3 and C-4, C-6 and C-7 C-12 and C-2, C-13 and C-14 (other couplings may have been missed due to poor incorporations).

[Cary, L. W., Seto, H., and Tanabe, M. (1973). *Chem. Commun*, 867.]

2.2. (difficult problem). The two metabolites [A] and [B] were isolated from *Periconia macrospinosa*: both were enriched with ^{13}C following incorporation of [^{13}C]-acetates. Labelling patterns were assigned following incorporation of [1-^{13}C]acetate (●) and [2-^{13}C]-acetate (*). In addition. ^{13}C-^{13}C coupling was observed following incorporation of [1,2-^{13}C]-acetate, and these data are shown below. Suggest biosynthetic pathways to both metabolites.

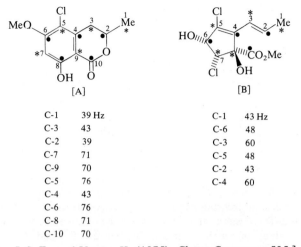

C-1	39 Hz		C-1	43 Hz
C-3	43		C-6	48
C-2	39		C-3	60
C-7	71		C-5	48
C-9	70		C-2	43
C-5	76		C-4	60
C-4	43			
C-6	76			
C-8	71			
C-10	70			

[Holker, J. S. E., and Young, K. (1975). *Chem. Commun.*, 525.]

2.3. The fungal product citromycetin incorporates label from acetate, and the results are shown below:

$$\text{Me}\overset{\bullet}{\text{C}}\text{O}_2\text{H} \longrightarrow$$

[13]C-nmr data (when [1,2-^{13}C]-acetate was employed)

Carbon	$J^{13}C-^{13}C$
1	51.3
2	51.1
3	56.2
4	56.4
5	47.9
6	48.3
7	67.4
8	67.3
9	72.5
10	72.7
11	70.2
12	70.6
13	72.0
14	72.0

How might this molecule be assembled?
[Evans, G. E. and Staunton, J. (1976). *Chem. Commun.*, 760.]

2.4. The xanthone, ravenelin, is produced by certain lower fungi. [1,2-^{13}C]-acetate is incorporated as shown. Propose a reasonable biosynthetic pathway.

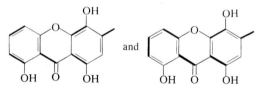

[Birch, A. J., Baldas, J., Hlubucek, J. R., Simpson, T. J., and Westerman, P. W. (1976). *J. Chem. Soc. Perkin Trans. I*, 898.]

2.5. (difficult problem). The pyrone shown below is produced by *Aspergillus melleus*, and incorporates [1,2-^{13}C]-acetate to provide the pattern illustrated. There was a small [13]C-^{13}C coupling observed between C-1 and C-7, as well as the much larger coupling between C-4 and C-5, C-2 and C-3, and C-8 and C-9. Suggest a biosynthetic route to this metabolite.

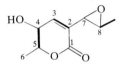

[Holker, J. S. E. and Simpson, T. J. (1975). *Tetrahedron Lett.*, 4693.]

2.6. Monocerin is produced by a variety of microorganisms, and has biological activity against powdery mildew of wheat. Identify a possible polyketide intermediate, and predict which carbon signals (^{13}C-resonances) would show isotope-induced shifts if [1-^{13}C, [2]H₃]-acetate and [1-^{13}C,[18]O₂]-acetate were used for incorporation studies.

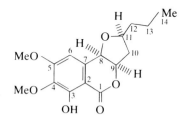

[Scott, F. E., Simpson, T. J., Trimble, L. A., and Vederas, J. C. (1984). *Chem. Commun.*, 756]

2.7. O-methylasparvenone has been isolated from *Aspergillus parvulus*, and incorporation studies with [1,2-^{13}C]-acetate provided the results which follow:

Carbon	$J\,^{13}C-^{13}C$
1	42
2	41
3	37
4	37
4a	62
5	63
6	70
7	71
8	61
8a	62
9	33
10	34

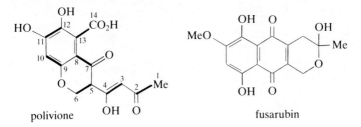

Propose a biogenesis for the metabolite, based upon this data.
[Simpson, T. J. and Stenzel, D. J. (1981). *Chem. Commun.*, 239.]

2.8. (difficult problem). Polivione is a major metabolite of *Penicillium frequentans*, and incorporates label from [1,2-^{13}C]-acetate to yield the pattern shown. When [1-^{13}C,^{18}O$_2$]-acetate was employed, C-7, C-9, and C-11 exhibited isotopically shifted ^{13}C-signals. When ^{18}O$_2$ was administered, C-4, C-12, and C-14 exhibited shifted signals. Given that the metabolite fusarubin probably has a related biogenesis (at least in the early stages), propose a biogenetic pathway to polivione (Hint: the presence of an ^{18}O-isotope attached to C-4 is especially significant.)

polivione   fusarubin

[Demetriadou, A. K., Laue, E. D., and Staunton, J. (1985). *Chem. Commun.*, 764]

3. Metabolites derived from mevalonate: isoprenoids

Metabolism of acetate also gives rise to a large and structurally very diverse group of secondary metabolites: the *isoprenoids* or terpenoids. Many of these compounds are familiar to us (at least by name), for example, menthol [85], camphor [86], cholesterol [87], and vitamins A [88] and D [89].

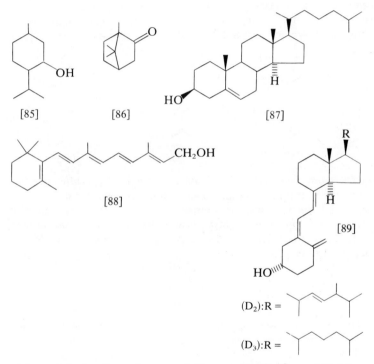

[85]

[86]

[87]

[88]

[89]

$(D_2):R =$

$(D_3):R =$

Many of the smaller and more volatile compounds (C_{10} and C_{15}) have been used by Man, albeit as components of crude plant extracts, for at least two thousand years. Certainly, any mediaeval herbalist worth his salt would have numerous recipes for the production of medicines and essences from rosemary, thyme, tansy, etc., all of which are rich in terpenes. Today, monoterpenes (C_{10}) and sesquiterpenes (C_{15}) are the basis of many of our finished goods: perfumes for cosmetics, soaps, and detergents; for flavouring certain processed meats and tinned goods; in confectionery; and as active ingredients of oral preparations, e.g. toothpaste and mouthwash. Tonne quantities of dried plants and fresh foliage are processed each year to produce the essential oils which are the raw materials of this industry.

It is interesting to note that our distant ancestors used these self-same plant extracts as perfumes (to dispel the odour of unwashed humanity) and as flavour enhancers (to obscure the taste of putrid meat, etc.).

The ready availability of these compounds together with their utility provided an early incentive for chemists to investigate them, and the structures of many of the simpler ones were established in the last century. It soon became apparent that the compounds possessed a common structural feature: they contained an integral number of C_5 units, as shown for geraniol [90], thujone [91], and camphor [86].

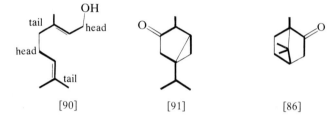

[90]   [91]   [86]

Furthermore, isoprene, 2-methylbuta-1,3-diene [92], was often obtained on pyrolysis of these C_{10} compounds, and it was suggested that isoprene was the building block for terpene biosynthesis: condensation of successive isoprene units in a head-to-tail fashion would produce compounds of formula $(C_5)_n$. This was known as 'the isoprene rule', and hence the term isoprenoids.

[92]

However, there were apparent exceptions to this rule, and to explain the biogenesis of irregular terpenoids it was necessary to think in terms of a 'biogenetic isoprene rule' (Ruzicka 1953). That is, terpenoids were assembled from C_5 units (isoprene-like), but the initially formed structure could be modified enzymatically in a number of ways to generate the known skeletal types. This is of course a common finding: secondary metabolic pathways yield a basic skeletal type which is then further elaborated.

Both of these rules are strictly empirical, but have been of immense utility in establishing the biogenetic origins of secondary metabolites.

Biosynthesis of the C_5 unit

The identity of the actual biological C_5 unit was in doubt for a considerable time. A number of C_5 acids and aldehydes occur naturally (Fig. 3.1), and were considered as possible precursors of the isoprenoids.

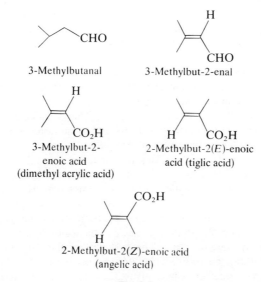

3-Methylbutanal

3-Methylbut-2-enal

3-Methylbut-2-
enoic acid
(dimethyl acrylic acid)

2-Methylbut-2(E)-enoic
acid (tiglic acid)

2-Methylbut-2(Z)-enoic acid
(angelic acid)

Fig. 3.1

Some of these compounds do in fact arise as side-products of the main polyisoprenoid pathway, but others, such as tiglic and angelic acids, derive from the metabolism of amino acids (isoleucine in this instance). The first major advance came with the discovery of mevalonic acid (MVA) [93], a component of the brewers' solubles which could support the growth of an

(3R)-MVA

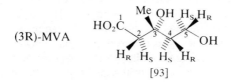

[93]

acetate-requiring mutant strain of bacteria, when acetate was omitted from the growth medium. Radioactive MVA((3R)-enantiomer) was incorporated almost quantitatively into cholesterol [87], with loss of carbon dioxide.

Subsequent elucidation of the biosynthetic pathway to MVA, and the mechanism of its conversion into the biological C_5 unit, and thence into isoprenoids, is one of the great sucess stories of the post-war years. It was due mainly to the pioneering work of Bloch, Lynen, Cornforth, and Popják, working with extracts from yeast or mammalian liver. These systems produce steroids (such as cholesterol) as their major metabolites, but the early stages of the isoprenoid pathway have been shown to be common to all organisms that have been studied, so the findings of Cornforth *et al.* are universally applicable.

The mevalonic acid molecule possesses a chiral centre at C-3, and only one of the two possible enantiomers (3R) is utilized in isoprenoid biosynthesis. In addition the molecule possesses six, prochiral, methylene hydrogen atoms. That is, replacement of any one of these atoms with a tritium (or deuterium) atom will produce a new chiral centre, and in the case of tritium, a radioactive molecule as well. This is depicted in structure [93], where H_R is a proton which, if replaced by an isotope of hydrogen, will confer the (R)-configuration on the carbon atom, i.e. it is pro-(R). Replacement of an H_s proton similarly produces a new chiral centre with (S)-configuration (pro-(S)). The fate of each methylene hydrogen atom in the biosynthesis of isoprenoids, has been determined by using each of the six possible, (3R)-monotritiated species of MVA as precursors in biosynthetic experiments. In this way it has been shown that each step on the pathway from MVA to the C_5 unit, and thence to the isoprenoids, is, not surprisingly, stereospecific. The first part of this process, which is common to all organisms producing isoprenoids, is shown in Fig. 3.2.

A double condensation (Claisen-type, then aldol) between molecules of acetyl-SCoA leads successively to acetoacetyl-SCoA [5], and 3-hydroxy-3-methyl-glutaryl-SCoA [94], HMG-CoA. There follows an essentially irreversible (probably rate-limiting) two-step reduction via hemithio-acetal [95], to produce (3R)-MVA [93]. These enzymic processes are subject to very complex regulatory controls, and there appear to be active and inactive forms of HMG-CoA reductase. Short-term regulation of, for example, steroid biosynthesis (*vide supra*) may thus be effected by interconversion of these forms, and when this process is better understood it should be possible to design drugs which would control cholesterol production. MVA is then phosphorylated, and MVA-5-pyrophosphate is decarboxylated and dehydrated, to yield isopentenyl pyrophosphate (IPP) [96]. 3-phospho-MVA-5-pyrophosphate is a likely intermediate in this process, since a mole of ATP is consumed: a 3-phosphate would also be a better leaving group than hydroxyl. IPP is then isomerized to dimethylallyl pyrophosphate (DMAPP) [97], via addition of a proton from the medium) and abstraction of the pro-(R)-hydrogen from C-2 (originally the 4-pro-(S)-hydrogen of MVA). The stereochemistry of this process is suggestive of a concerted mechanism, and it is reasonable to assume the isomerization occurs as depicted in eqn (3.1). The isomerization is formally

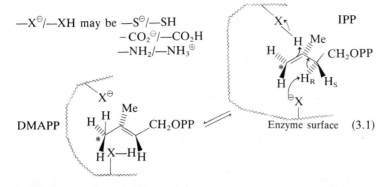

reversible though essentially irreversible in many organisms: an equilibrium ratio of IPP:DMAPP of 7:93 was obtained with a yeast extract. However, even if the two C_5 units are interconverted several times before they are utilized, the two methyl groups of DMAPP do not as a rule become equivalent (note * in Fig. 3.2 and in eqn 3.1), such is the stereospecificity of the isomerase enzyme.

These two activated C_5 units are the biological equivalents of isoprene, so-called 'active isoprene', and isoprenoids are the products of their union. DMAPP is an alkylating agent *par excellence* since it can suffer nucleophilic attack at C-1 with concomitant loss of pyrophosphate which is a good leaving group. The currently available evidence would suggest that an S_N1-type process is involved, albeit with an allyl cation: pyrophosphate ion pair that is highly structured (Fig. 3.3). The rates of condensation of various modified substrates (also shown in Fig. 3.3) with IPP catalysed by pig-liver enzymes were shown to be much lower than with the natural substrates (DMAPP and geranyl pyrophosphate, respectively). These rate reductions parallel those observed when the corresponding methane sulphonates undergo solvolysis under typical S_N1 conditions. Such rate reductions would be predicted on the basis of the potent destabilizing effect of a fluorine and trifluoromethyl group ($-I$ effects) on proximal carbocations.

One obvious objection to these results would be that the modified substrates are not true substrates, but at least in the case of 2-fluorogeranyl pyrophosphate this cannot be sustained since this substrate yields the anticipated product 6-fluorofarnesylpyrophosphate. Finally, elegant experiments using ^{18}O-labelled substrates have established that a highly structured ion pair is involved, such that scrambling of the label does not occur prior to reaction with IPP; this is also shown in Fig. 3.3.

The stereochemistry of these processes is well established, and the 2-pro-(R)-hydrogen of IPP is lost to form a new *trans* double bond. Alternatively, the 2-pro-(S)-hydrogen is lost when forming a new *cis*

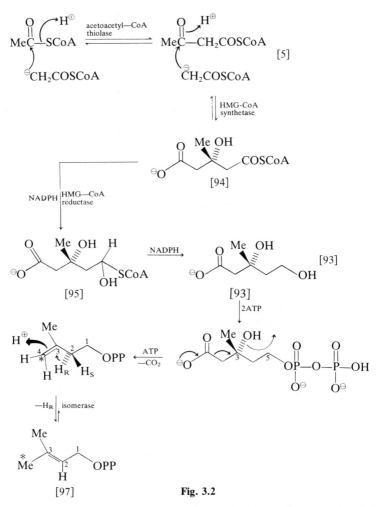

Fig. 3.2

double bond, as in the biosynthesis of natural rubbers. The stereochemical outcome of the reactions is almost certainly a consequence of different conformations in which IPP may be held at the active site of the enzyme (or enzymes) responsible: prenyl transferase (s).

Almost any number of C_5 units may be incorporated to produce the diverse array of isoprenoids encountered in Nature. Most of these have all-*trans* stereochemistry and the main structural types are shown in Fig. 3.4. (The mechanism of formation of the tri- and tetraterpenes is different from that depicted in Fig. 3.3, and will be discussed in due course.)

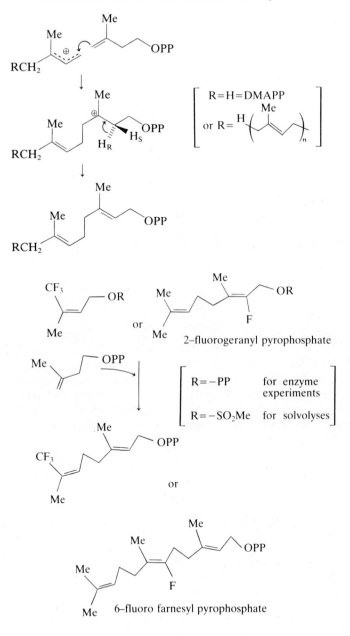

Fig. 3.3

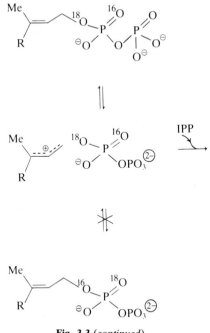

Fig. 3.3 (*continued*)

About 1 per cent of all plant species have the capacity to synthesize *cis*-polyisoprenoids, and the most important of these compounds is rubber [98]. The commercial source of rubber, *Hevea brasiliensis* has been commerically selected for its capacity to convert available MVA almost exclusively into rubber.

$$GPP(C_{10}) + n\,IPP \longrightarrow$$

$$n = 500 - 5\,000$$

[98]

Finally, a number of isoprenoids possess a mixture of *Z*- and *E*-double bonds, although there tends to be a predominance of the latter type. They occur primarily in non-photosynthetic plant tissue, and are probably important components of various membraneous structures.

In the following sections we shall consider the various classes of terpenoids in more detail.

Hemiterpenes

Since DMAPP [97] is such an effective electrophile, it is not surprising that the 3,3-dimethylallyl-group is present in many secondary metabolites

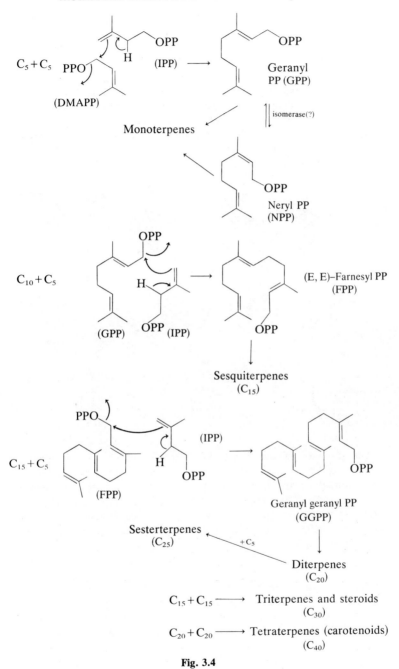

Fig. 3.4

which derive the major part of their skeletons via non-mevalonoid pathways. Products of both C- and O-alkylation are common, and the dimethylallyl-group is invariably located as sites which were incipiently nucleophilic prior to alkylation, such as active methylenes and phenolic oxygen atoms.

Some typical examples of these hemiterpenes are humulone [99], a major constituent of hops, and responsible for much of the bitter flavour of beer; echinulin [100], a fungal metabolite which derives the major part of its skeleton from the amino acids, tryptophan and alanine; and the ergot alkaloids such as lysergic acid [101] (the diethylamide of this acid is the potent hallucinogen LSD).

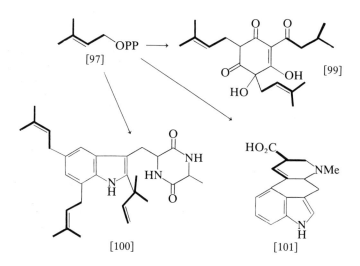

The presence of the dimethylallyl-unit is not always immediately obvious, as for example in lysergic acid, and in the case of certain furan-containing compounds, e.g. [102]; it is obscured by the fact that the C_5 unit suffers degradative modification after insertion. The co-occurrence of precursor species which retain the dimethylallyl-group, albeit in modified form, provided the clue which led to elucidation of the pathway shown in eqn (3.2).

Some of these hemiterpenes (furanocoumarins, e.g. [102], and ergot alkaloids) will be discussed in more detail in Chapter 6, which is concerned with secondary metabolites whose skeletons incorporate structural units derived from a variety of pathways, i.e. products of mixed metabolism. Finally, isoprene itself, [92], is emitted from the leaves of certain plants, and several C_5 alcohols derived from IPP and DMAPP also occur naturally.

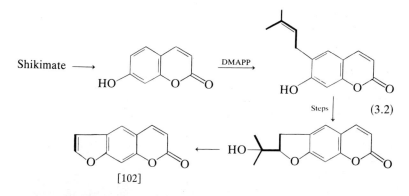

Shikimate →

$$(3.2)$$

[102]

Regular monoterpenes

Investigation of monoterpene biosynthesis has always been hampered by poor incorporation of radiolabelled MVA by whole plants. Often as little as 0.01 to 0.1 per cent of the radiolabel is found in the isolated monoterpenes. However, somewhat better incorporations have been achieved recently using cell-free systems, and there is no doubt that MVA is an obligate precursor of the monoterpenes.

Alicyclic, monocyclic, and bicyclic compounds are found in many plants, and in certain insects, but only rarely in animals. Enzyme studies have shown that both acyclic and cyclic compounds can be derived from geranyl PP [103] and from neryl PP [104] (GPP and NPP, see Fig. 3.5). A fundamental problem arises here: since neryl PP has a 2-Z-double bond, if formed directly from IPP and DMAPP, two different modes of coupling must be invoked to explain the formation of geranyl PP (2-E) and neryl PP (2-Z). Possible reaction modes are shown in eqn (3.3) (cf. Fig. 3.3). However, tracer studies using (4R)- and [4S)-4-³H₁]-MVA (i.e. monotri-

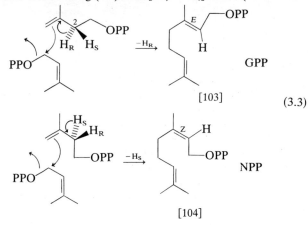

$$(3.3)$$

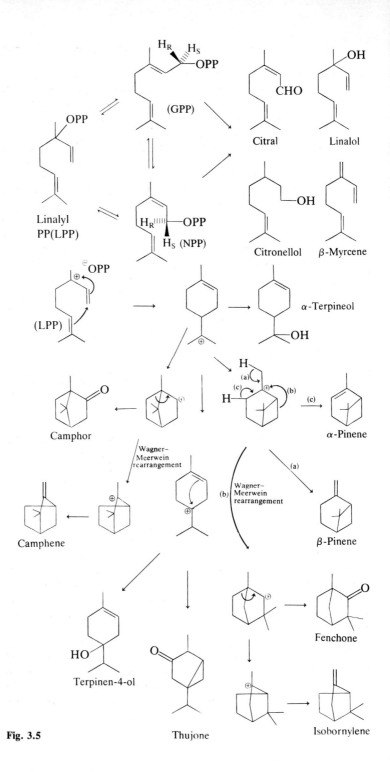

Fig. 3.5

ated MVA species) showed that the (4S)-hydrogen was lost in both the formation of geraniol (the alcohol derived from [103], see [90] and of nerol (the alcohol from [104]). (Note that the 4-pro-(S)-hydrogen if MVA becomes the 2-pro-(R)-hydrogen of IPP.) This evidence is in accord with either a common pathway, whereby GPP is first formed and then isomerized (perhaps via the aldehyde) to NPP; or with the existence of two separate transferases (or perhaps one enzyme with two discrete active sites) which catalyse the formation of either GPP or NPP. There is a certain amount of evidence in support of both of these possibilities. For example, Banthorpe has demonstrated that a stereospecific isomerization occurs in intact *Rosa damascena* by using [^{14}C, ^{3}H]-doubly labelled species. Geraniol was converted into nerol with loss of the 1-pro-(S)-hydrogen, while the reverse process involved loss of the 1-pro-(R)-hydrogen from neryl pyrophosphate. It is thus reasonable to implicate redox processes with interconversion taking place via the respective aldehydes. Cell-free extracts from the same species required a supply of the cofactor couple NADP$^{\oplus}$/NAPDH in order to carry out this isomerization.

In contrast, enzyme systems from *Salvia officianalis* (sage), *Tanacetum vulgare* (tansy), *Mentha spicata* (spearmint), and *Foeniculum vulgare* (fennel), amongst others, all produce cyclic monoterpenes from geranyl pyrophosphate, neryl pyrophosphate, but best of all, from linalyl pyrophosphate. In consequence, another reasonable pathway would be GPP ⇋ LPP ⇋ NPP, and this is also shown in Fig. 3.5.

On balance, it appears that the cyclase enzymes actually function as isomerase-cyclases, and take geranyl pyrophosphate through to linalyl pyrophosphate (or the ion-pair equivalents—see Fig. 3.6 for examples), and that LPP is then the major precursor of the cyclic monoterpenes.

It was originally suggested that cyclization to produce monocyclic and bicyclic monoterpenes proceeds via cationic intermediates, with re-arrangements of the Wagner–Meerwein type, and hydride shifts, giving rise to the diverse array of skeletal types that have been found in Nature. A hypothetical biogenetic scheme, based upon one proposed by Ruzicka, is shown in Fig. 3.5. Certainly the chemical acrobatics of the monoterpenes are well-known *in vitro* (e.g. the rearrangements of the bornyl skeleton and the classical/non-classical carbonium ion controversy). It is most unlikely that discrete cationic species exist *in vivo*, but rather a series of tight ion pairs with pyrophosphate anion are probably involved. This has been established in the case of 2-bornylpyrophosphate produced by sage (Fig. 3.6). In addition, the conventional flat structures shown in Fig. 3.5 do not provide good representations of the actual conformations of LPP involved in the cyclizations; rather better representations are given in Fig. 3.6. These are due to Cane and Croteau who have carried out much of the recent work involving enzyme systems responsible for cyclization.

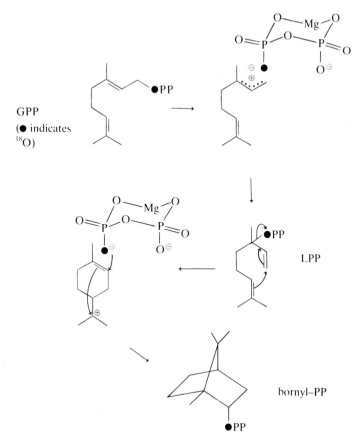

GPP
(● indicates
^{18}O)

LPP

bornyl–PP

Fig. 3.6

In the few instances where complete degradation of labelled monoterpenes has been accomplished, it has been found, somewhat surprisingly, that label from MVA or from $^{14}CO_2$ (supplied at normal physiological concentrations) resides almost entirely in the upper halves of the molecules. That is, label is incorporated from IPP, but not, apart from a few per cent of the total incorporated radioactivity, from DMAPP. Thus for thujone [91], from *Tanacetum vulgare* (tansy), label was only found at the carbonyl carbon (eqn 3.4). Similar results were obtained with various pinus species after feeding [2-^{14}C, 2-^{3}H$_2$]-MVA, and the pathways shown in eqn (3.5) have been proposed in order to account for the predominance of ($-$)-β-pinene and ($+$)-α-pinene in most pinus species. The less common stereoisomers probably arise by isomerization.

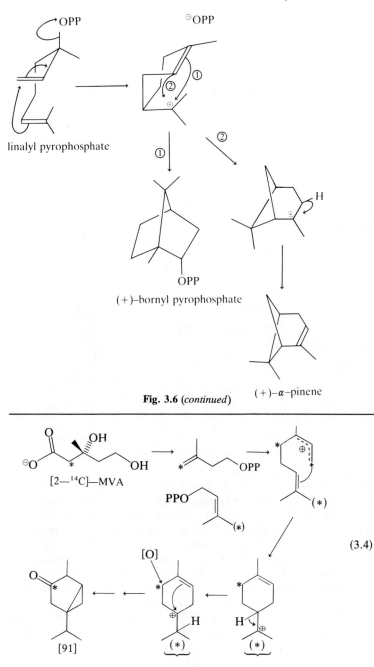

linalyl pyrophosphate

(+)-bornyl pyrophosphate

Fig. 3.6 (*continued*)

(+)-α-pinene

[2—¹⁴C]—MVA

PPO

[O]

[91]

(3.4)

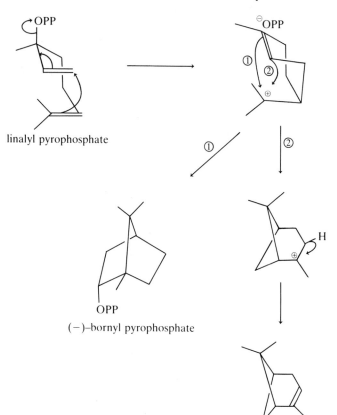

linalyl pyrophosphate

(−)-bornyl pyrophosphate

(−)-α-pinene

Fig. 3.6 (*continued*)

This appears to be a general phenomenon (similar asymmetric labelling patterns have been obtained for camphor [86], other monoterpenes and several sesquiterpenes), and various explanations have been considered. It is most likely that in some plants a 'pool' of DMAPP exists, and labelled IPP reacts with this inactive DMAPP before it (IPP) has time to isomerize to yield radioactive DMAPP. The source of this endogenous DMAPP is not known, but a number of metabolic pathways produce this metabolite. It may also be derived from degradation of monoterpenes since these are now known to be metabolically active, and often have a high rate of turnover: they are interconverted and degraded.

The known exceptions to this 'asymmetric' mode of biosynthesis are mostly conjugated monoterpenes, that is, monoterpene alcohols which occur as O-glucosides (an ether link exists between alcohol and glucose).

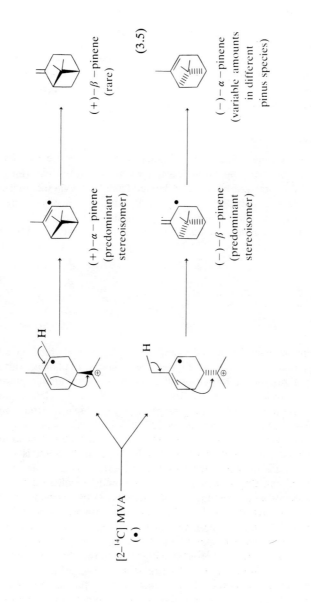

(3.5)

Thus, geranyl and neryl-β-D-glucosides, which occur in rose petals, are labelled in both halves of the monoterpene moieties, when [2-^{14}C]-MVA is precursor. It is tempting to suggest that this 'symmetric' pathway represents a special type of monoterpene metabolism, especially since these glucosides occur almost exclusively in petals (which differ structurally from leaf tissue), whilst non-conjugated monoterpenes are present in foliage. Geranyl and neryl glucosides may in fact be storage compounds, acting as potential precursors for monoterpene interconversion and degradation. A definitive explanation is not available at present.

Irregular monoterpenes

It was mentioned earlier that irregular monoterpenes are encountered in Nature, and are apparent exceptions to the isoprene rule. In all probability, most, if not all, are derived from MVA via the intermediacy of the two C_5 units, IPP and DMAPP, but subsequent rearrangements give rise to structures which are not obviously isoprenoid in character. The three most common skeletal types are chrysanthemyl [105], artemisyl [106], and santolinyl [107]. Compounds of these three structural types

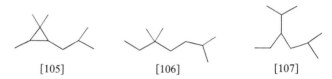

[105]   [106]   [107]

are almost exclusive to the Compositae (the family which includes the daisy), and since two skeletal types often co-occur in plants of this family, it seems reasonable to propose that a common biosynthetic precursor is involved. Indeed, *in vitro* studies have shown that the skeletal types are to some extent interconvertible, and could conceivably arise from a common cationic species, as shown in Fig. 3.7. It is interesting that in one model (i.e. non-enzymatic) experiment, a few per cent of a 'head-to-head' compound [108] was obtained, as this type of skeleton is of pivotal importance is the biosynthesis of tri- and tetraterpenes.

The biosynthetic route to these compounds is still in doubt, though experiments which employed cell-free extracts of *Artemesia annua* and *Santolina chamaecyparissus*, demonstrated the involvement of an enzyme sulphydryl group (sulphonium ion intermediate?), and that IPP and DMAPP were incorporated, but geranyl- and neryl-pyrophosphates were not. A scheme was proposed by Banthorpe and Doonan which incorporates the results of these experiments (Fig. 3.8).

One class of chrysanthemyl metabolites, the pyrethrins, are of particular interest. These metabolites have marked insecticidal activity, and may confer some ecological advantage on the plants which produce

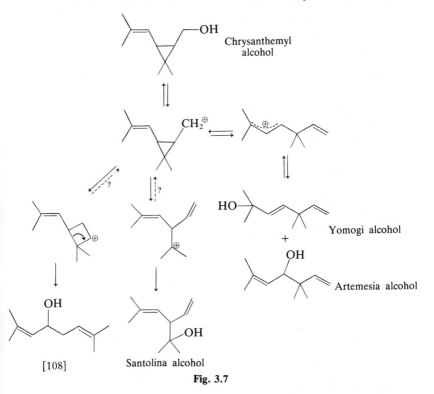

Fig. 3.7

them. The natural compounds are of the type [109], while more potent synthetic analogues incorporate furan rings or 3-phenoxybenzyl moieties [110] in order to render the compounds more fat-soluble; analogues with a *cis*-disposition of ester and alkylidene groups are even more potent, e.g. [111]. The synthetic pyrethrins are used in many domestic insecticidal sprays, and like the natural pyrethrins they are relatively non-toxic to animals, and biodegradable: two important attributes of the ideal insecticide. They are, however, very toxic to fish, and their use near rivers and streams (or near goldfish bowls!) is precluded for this reason.

The iridoids

The iridoids are a group of monoterpene metabolites characterized by skeletons in which a six-membered ring containing an oxygen atom is fused to a cyclopentane ring. Some typical examples are iridomyrmecin [112], iridodial [113], loganin [114], and nepetalactone [115]. The term 'iridoid' is derived from the name of the species of ants (*Iridomyrmex*) which utilize [112] and [113] in defensive secretions: the compounds are

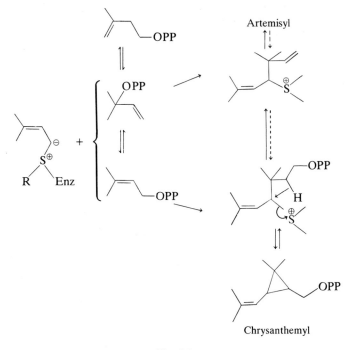

Fig. 3.8

irritants and deter most predators. Nepetalactone, from the catnip plant, is also an insect repellent, and has in addition an excitatory effect on cats, possibly because it has an odour reminiscent of their marker scent. Almost all irridoids are of plant origin, and usually occur as their glucosides.

Although it is not perhaps immediately obvious that these metabolites are derived from MVA, tracer experiments have proved this beyond doubt. The biogenesis of the basic skeleton is believed to be as shown in Fig. 3.9. Both hydroxygeraniol [116] and hydroxynerol [117] are efficient precursors, as are the corresponding dialdehydes (R = Me); but it is not yet clear whether trialdehydes (R = CHO) are also on the pathway, or whether oxidation occurs at a later stage. Although the mode of ring closure is not known, the mechanism depicted in Fig. 3.9 (proposed by Arigoni and Battersby) is reasonable in the light of tracer results.

Apart from their obvious interest as natural insecticides, the iridoids, and in particular loganin [114], have attracted much attention since they give rise to other important metabolites. Loganin is a key intermediate on the biosynthetic pathway to the complex indole alkaloids, and these will be discussed in Chapter 6. Suffice to state at this point, that loganin is

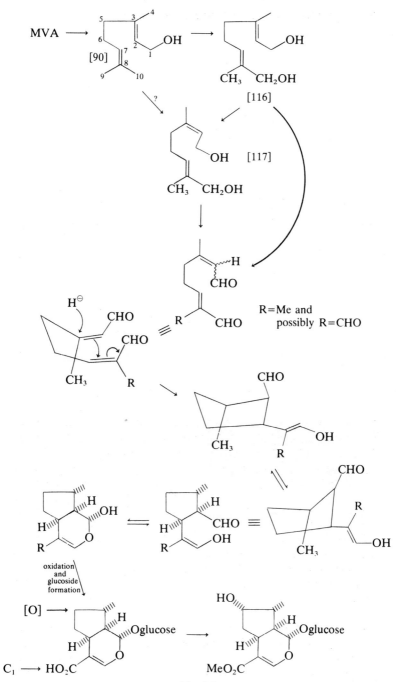

Fig. 3.9

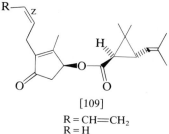

[109]

R = CH=CH₂
R = H

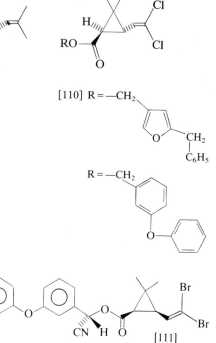

[110] R = —CH₂

R = —CH₂

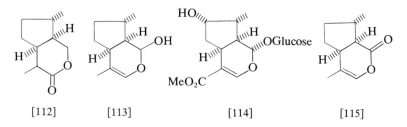

[111]

converted into secologanin [118], and thence (by condensation with the amine tryptamine) into isovincoside [119], the immediate precursor of the indole alkaloids. The mechanism of ring cleavage that produces secologanin is unknown, but cleavage conceivably occurs after prior oxidation of the methyl group, and formation of a phosphate ester. Secologanin is also the parent compound of a group of metabolites known as *secoiridoids*, of which gentiopicroside [120] and sweroside [121] are representative examples. These metabolic pathways are shown in Fig. 3.10.

[112]   [113]   [114]   [115]

Sesquiterpenes

As a family, the sesquiterpenes encompass an almost bewildering array of
structural types and more than fifty basic skeletons have been recognized.
Each of these skeletal types can be assumed to derive from farnesyl
pyrophosphate (FPP), 2*E*,6*E*- [122] or from the 2*Z*,6*E*-isomer [123],
or from nerolidyl pyrophosphate [124], and intermediate cationic species
(cf. monoterpene biogenesis) have been invoked to explain how the
various structures arise. However, the stereochemistry of bond scission,
ring closure, and of the migration of hydrogens and alkyl groups, is almost
invariably that anticipated (albeit modified as to position of attack
by steric effects) for concerted processes (E2, S_N2 and S_N2'): i.e.
anti-stereochemistry. It is reasonable to assume that such stereospecificity

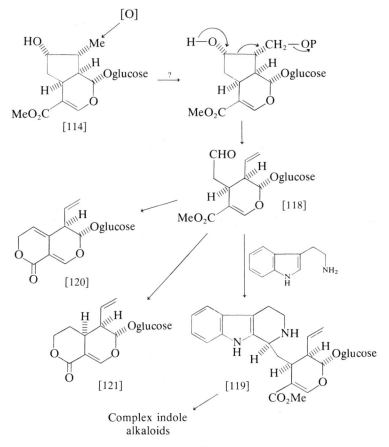

Fig. 3.10

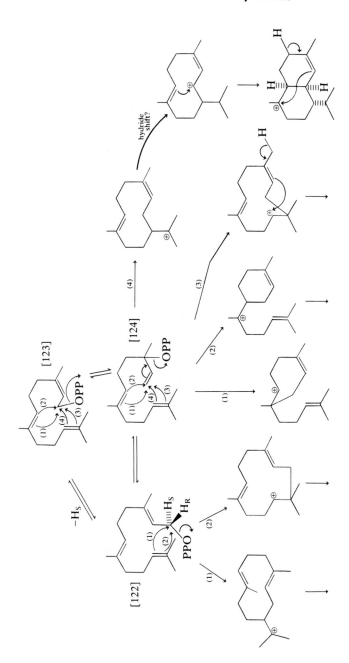

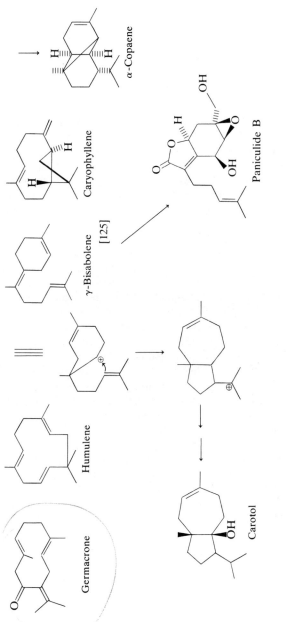

α-Copaene

Caryophyllene

γ-Bisabolene

[125]

Paniculide B

Humulene

Germacrone

Carotol

Fig. 3.11

is observed because the enzymes(s) act as a template for the acyclic precursor, allowing it to adopt the conformation from which the series of stereospecific transformations proceed with greatest facility. It is nevertheless useful to consider the hypothetical cationic intermediates, since they assist in the generation of biogenetic pathways. Indeed, many total syntheses of sesquiterpenes have been modelled on proposed pathways (i.e. biomimetic syntheses), and often proceed via intermediates which contain incipient cationic sites. Some reasonable pathways are shown in Fig. 3.11.

The figure is somewhat simplified and only one metabolite is shown emanating from each cationic species, but it does demonstrate the way in which the origins of the various skeletal types can be rationalized, at least on paper. Each cationic species can of course undergo rearrangement, or suffer hydride shifts, and functional groups may be introduced or modified. In some cases more than one route is possible, and stereochemical and energetic considerations may suggest which one is the more likely. A large number of elegant biosynthetic experiments have been carried out in the last ten years, notably by Arigoni, Hanson, Cane, and Overton, and we can now give credence to many of the pathways shown in Fig. 3.11.

For example, using tissue cultures of *Andrographis paniculata*, Overton has established that γ-bisabolene [125] is formed from 2*Z*,6*E*-FPP [123], or possibly from nerolidyl pyrophosphate [124]. The isomerization of [122] to [123] seems to proceed via two distinct mechanisms which are analogous to those proposed for the interrelationships of geranyl-, neryl-, and linalyl-pyrophosphates (Fig. 3.5). Thus the results of tritium-labelling experiments have shown that one pathway involves loss of the 1-pro-(S)-hydrogen of [122], while the other pathway proceeds from [122] via [124] to [123] without loss of either of the hydrogens at C-1.

In this same system they were able to demonstrate the anticipated labelling pattern for paniculide derived (via γ-bisabolene) from both [2-14C]-MVA and from [1,2-13C]-acetate.

In contrast to these results concerning the isomerization of farnesyl pyrophosphate, most other studies using cell-free systems have shown that the pathway via nerolidyl pyrophosphate is the most likely route for those sequiterpenes that require (at least formally) the 2*Z*,6*E*-FPP stereochemistry. The enzymes thus function as isomerase-cyclases. Once again, the conventional flat structures provide poor representations of the

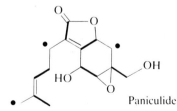

ex. [1,2–13C]–acetate

ex. [2–14C]–MVA

Paniculide

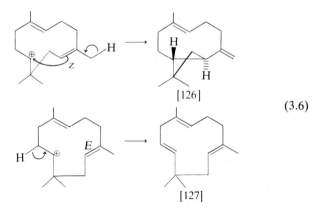

(3.6)

conformations involved in the cyclizations (though it has to be admitted that they are much easier to draw, and will be used for most of the figures), and more realistic structures are shown (again these are due to Cane and Croteau) for the case of trichodiene biosynthesis (see Fig. 3.14).

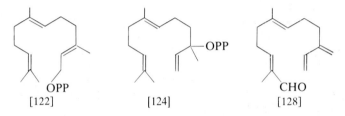

Before we pass on to consider some sesquiterpenes in more detail, brief mention should be made of the chemotaxonomic importance of these metabolites. Often different skeletal types co-occur in several species of a particular genus; and in some instances a particular species has been assigned to a genus because it produces metabolites with skeletons characteristic of that genus. Co-occurrence also suggests that a common biosynthetic progenitor is involved. That is, an initial pathway is shared by all of the metabolites, and there is a branching point and pathways leading from the common progenitor to the various skeletal types. Thus, the common intermediacy of γ-bisabolene [125] has been invoked to explain the frequent co-occurrence of four skeletal types (Fig. 3.12). Such schemes must, however, be regarded as hypothetical, and co-occurrence is a necessary rather than sufficient requirement for shared metabolic pathways. Caryophyllene [126] and humulene [127] often co-occur, but cannot have a common precursor on stereochemical grounds (eqn. 3.6).

Acyclic sesquiterpenes

Only relatively few acyclic sesquiterpenes occur naturally, and all are derived from farnesol, the alcohol of FPP [122], nerolidol, or from their

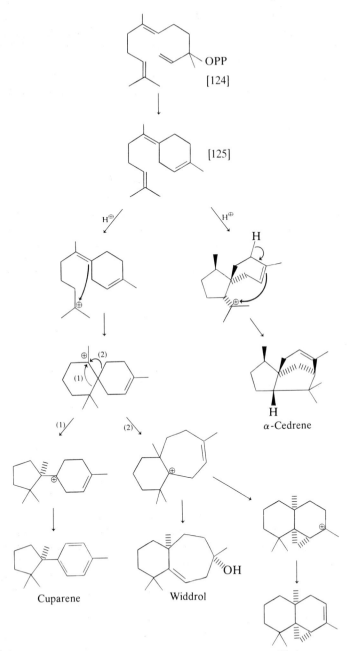

Fig. 3.12

pyrophosphate esters [122] and [124] which occur widely. With the exception of the farnesenes (various hydrocarbons derived from [122] and [124]), and one or two other metabolites, e.g. β-sinensal [128] (a flavour constituent of Chinese oranges), most of the known acyclic sesquiterpenes contain furan rings, or tetrahydrofuranyl moieties. Representative examples are davanone [129], freelingyne [130]

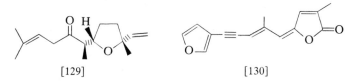

[129]   [130]

(acetylenic isoprenoids are very rare), and ipomearone [131]. The latter is of considerable interest since it is a fungicide produced by potatoes (and a few other plants) in response to fungal infection by a black-rot fungus. It is not produced by healthy potatoes. Such antimicrobial metabolites, produced only in the diseased state, are known as *phytoallexins*. Several tracer experiments have been conducted: [2-^{14}C]-MVA and

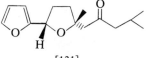

[131]

[2-^{14}C]-farnesol are incorporated into ipomearone; and there is little doubt that these compounds are isoprenoids.

In many ways the most interesting member of this class is the so-called juvenile hormone [132a] of the giant silkworm moth, *Hyalophora cecropia*, which also occurs in small amounts in other insect species.

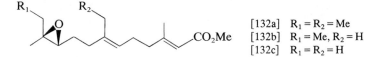

[132a] $R_1 = R_2 = Me$
[132b] $R_1 = Me, R_2 = H$
[132c] $R_1 = R_2 = H$

The physiological effect of [132a] and co-occurring [132b] is to control the maturation of the insect larva. While juvenile hormone is present no metamorphosis into an adult insect occurs, and successive larval moults take place instead. Under natural conditions, production of the hormone ceases at a particular stage, and metamorphosis occurs. Addition of synthetic juvenile hormone (and certain analogues) causes continual moulting with no metamorphosis (a giant larva is produced), and in consequence these compounds have potential as insecticides.

The skeleton is C_{15} with one or two extra methyl groups (R_1, R_2), and it is tempting to believe that juvenile hormone is derived from FPP with the addition of two methyl groups from S-adenosyl methionine. However, although the methyl group of the ester functionality is derived from the coenzyme, the other methyl groups are not. Some recent incorporation studies have helped to clarify the situation. [2-¹⁴C]-MVA was incorporated into [132b] and [132c] by gland cultures of another insect species and it was shown that while [132c] is derived from three moles of MVA, [132b] incorporates two moles only. By inference, [132a] should incorporate only one mole of MVA. In addition, [132c] incorporates nine moles of acetate, while [132b] incorporates eight moles of acetate and one of propionate. Finally, [1-¹⁴C]-propionate was incorporated specifically into [132a] by Cecropia, and a feasible pathway which takes into account all of this information, is shown in eqn (3.7).

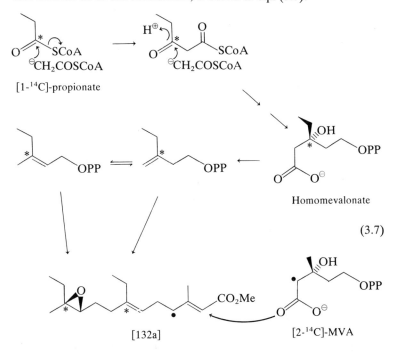

(3.7)

Several juvenile hormone analogues with broad spectrum insecticidal activity have been synthesized. Such compounds are rapidly biodegradable (often before they have a chance to act!), and along with the pyrethrins [109]–[111] are potentially ideal insecticides, though initial results from field trials with JH analogues have been disppointing.

Cyclic sesquiterpenes
It would be impossible to discuss here even a fraction of the known compounds which comprise this class of metabolites: many excellent

reviews and books are available, and we shall confine our attention to metabolites of particular contemporary interest.

As in the case of the acyclic compounds, most of the biologically active species are oxygenated, and it is these compounds which have received the most attention. Many have skeletons which can only arise by extensive rearrangement and bond cleavage of hypothetical cationic intermediates. Most of these pathways are almost unique, shared by a very small number of similar metabolites.

The fungal antibiotic ovalicin [133] has been the subject of recent definitive ^{13}C- and ^{2}H-n.m.r. studies, and is almost certainly derived via the pathway shown in Fig. 3.13. The structurally similar antimicrobial compound fumagillin [134] presumably arises by a similar route.

Another group of metabolites that are derived from 2E,6E-farnesyl PP [122] are the tricothecane antibiotics. These are, incidentally, very toxic, causing haemorrhage in mammals; and since they could conceivably be minor contaminants of many foodstufs, the long-term toxic effects are being studied. Incorporations of isotopic label from three stereospecifically tritiated MVA species, from [3-^{3}H$_{1}$]-GPP, from two tritiated FPP species, and most recently, from [2-^{13}C]-MVA, have been achieved. The results obtained are in accord with the pathway shown in Fig. 3.14. This incorporates a series of supposedly concerted 1,2-shifts, culminating in a 1,4-hydride shift to give the fungal metabolite trichodiene [135], which suffers oxidative modification to produce finally trichothecolone [136] (R = H), and trichothecin [136] (R = CO—CH=CHMe). Once again it was shown that [122] is converted into sesquiterpene (in this case, trichodiene) without loss of tritium from C-1 of tritiated [122], probably with the intermediacy of nerolidyl pyrophosphate [124]. Several of the trichothecanes possess marked antitumour activity, and both the natural compounds and synthetic analogues are being evaluated for possible clinical use.

The mushroom *Clitocybe illudens* (otherwise known as the Jack O'Lantern mushroom because of its bioluminescence) produces two major toxic sesquiterpenes: illudin M [137] (R = Me), and illudin S [137] (R = CH$_2$OH). A plausible pathway which involves the intermediacy of a cyclopropyl carbonium ion has been suggested to account for the tracer results obtained by Hanson and coworkers and this is shown in Fig. 3.15.

Finally, we should consider the large group of sesquiterpene lactones, which are of special interest since many of them possess potent anti-tumour properties. Once again, most of these metabolites occur in just one family of plants, the Compositae. Some typical examples are α-santonin [138], and the three, very potent anti-tumour agents, vernolepin [139], plenolin [140], and elephantin [141]. The modes of biogenesis are generally unknown with the exception of santonin, which is believed to derive from eudesmol [142]. It is known that lactone

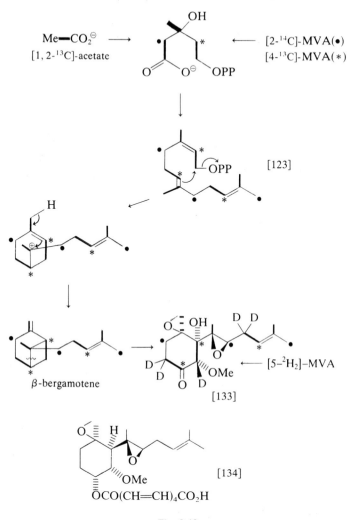

Fig. 3.13

formation precedes formation of the dienone grouping, and metabolites which resemble all of the proposed intermediates have been isolated, i.e. costal [143], costic acid [144], and costunolide [145].

The marked biological activity of these lactones, or more particularly, of α,β-unsaturated lactones derived from them, is thought to be due to their properties as Michael acceptors, that is as potent alkylating agents. Most strong alkylating agents are cytotoxic, (including perhaps DMAPP)

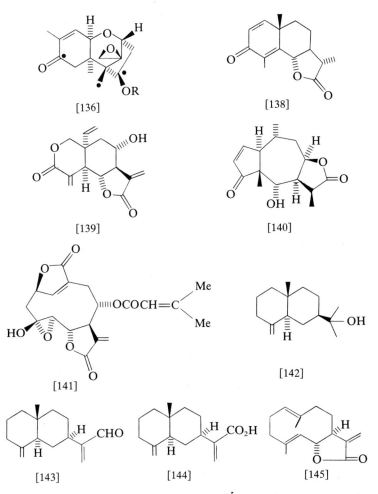

[136]   [138]

[139]   [140]

[141]   [142]

[143]   [144]   [145]

and there is no reason to suppose that the marked antitumour properties of the sesquiterpene lactones cannot be explained in the same way. Whether the powerful germination stimulant strigol [146] acts by this mechanism is open to speculation.

The number of sesquiterpene skeletal types is seemingly endless, and many new and more exotic metabolites are unearthed each year. In particular, marine organisms are proving to be a rich source of novel halogenated sesquiterpenes (and other halogenated species), such as the compound [147], from the red algo *Laurencia pacifica*. The one unifying factor is that all of these diverse skeletal types derive from farnesyl pyrophosphate.

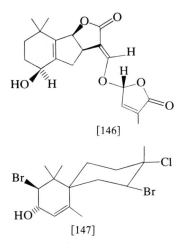

[146]

[147]

The rationalization of the diverse C_{15} structures is one of the most significant intellectual achievements in organic chemistry, certainly comparable with other, rather better-known advances like the Woodward–Hoffmann Rules.

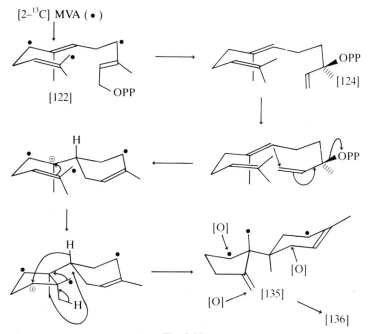

Fig. 3.14

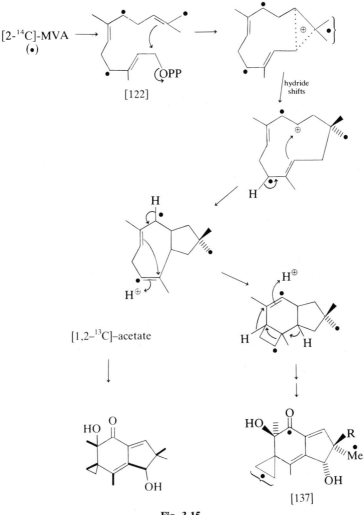

Fig. 3.15

Natural fragrances and flavours

Before we pass on to consider the higher terpenoids, a brief discussion of natural fragrances and flavours is worthwhile. We have already noted the commercial importance of the essential oils. Some of the most highly prized oils are: oil of lavender (linalol [148] is the major odiferous constituent), oil of lemongrass (2-Z-citral [149]), oil of sandalwood (the santalenes [150]), oil of patchouli (patchouli alcohol [151]) and attar of roses (geraniol [90] and citronellol [152]).

It is obviously nonsensical to suppose that the secondary metabolic pathways leading to the volatile mono- and sesquiterpenes have evolved for the benefit of Man's olfactory and gustatory senses, and it is thus important to consider the ecological significance of these compounds. This subject will be discussed in some detail in Chapter 7, but some of the principles will be mentioned here.

In the wild, any organism seeks to eat, escape the attentions of its predators, and reproduce; and in this endeavour it is in competition with members of its own species, and with the members of many alien species. At the simplest level, some plants attract insects as prey or to elicit their help in pollination, while other plants actively repel insects and other herbivores. Similarly, insects attempt to repel predators but attract mates.

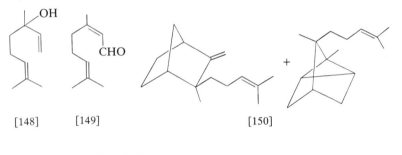

[148]   [149]   [150]

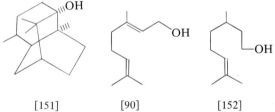

[151]   [90]   [152]

Since a volatile chemical species can be transmitted through space, in many instances plants and insects 'communicate' with the help of volatile terpenes. Thus, the common honeybee (*Apis mellifera*) emits a mixture of 2-Z-citral [149], nerolic acid [153], geraniol [90], and geranic acid [154], in order to attract other bees to a prime source of nectar which it has discovered.

The male cotton-boll weevil emits four rather peculiar monoterpene metabolites, [155]–[158], in order to attract the female beetle for mating. This follows his discovery of a suitable breeding site (i.e. a cotton plant), to which he was attracted by the odour of α-pinene [159] and other monoterpenes, released by the plant. In all probability he ingests this

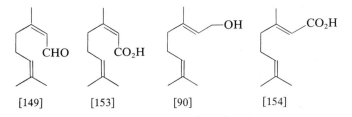

[149]   [153]   [90]   [154]

plant metabolite during an initial feeding period, and then synthesizes attracting compounds [155]–[158] from α-pinene, or some other monoterpene.

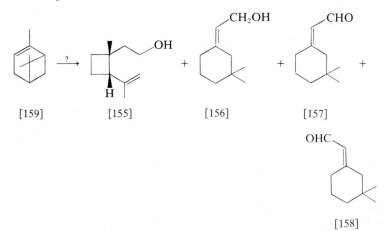

[159]   [155]   [156]   [157]   +

[158]

We have already seen that the iridoids, iridomyrmecin [112], and iridodial [113], are present in the defensive secretions of certain ants, and there are many other examples of volatile terpenes being used in a defensive role.

Here then are examples of three kinds of communication mediated by terpenes: trail laying, sexual attraction, and deterrence. The list of known ecological interactions mediated by volatile terpenes and other secondary metabolites is ever growing, and surely this area provides vast scope for fruitful collaboration between chemists and biologists.

Diterpenes

The progenitor of the C_{20} polyisoprenoids is almost certainly geranyl geranyl PP [160], or in some cases its isomer geranyl linalyl PP [161]. Acyclic metabolites are not common, although the monounsaturated diterpene phytol [162] is incorporated in the chlorophyll molecule, and is thus ubiquitous. Indeed, phytol, and phytane (the fully saturated hydrocarbon derived from it), are present in lake and river sediments, both ancient and modern: their presence in ancient sediments is good evidence for the antiquity of both the chlorophyll molecule and the mevalonate pathway.

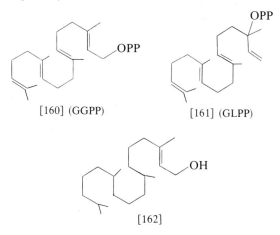

Many different monocyclic and polycyclic metabolites are encountered, and acid-catalysed, stereospecific cyclizations of geranylgeraniol (the alcohol of [160]) and analogues give rise to a variety of diterpene skeleta. A similar process may be assumed to occur *in vivo*. Two possible modes of cyclization are possible, leading to a $5\alpha,10\beta$ configuration (most commonly encountered) [163], or to a $5\beta,10\alpha$ configuration [164] at

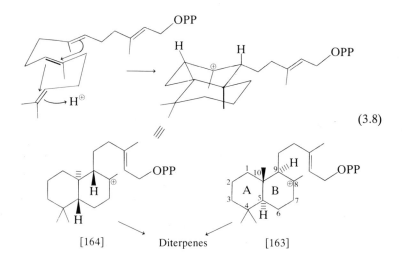

(3.8)

the A:B ring junction (eqn 3.8). Diterpenes are derived via further modification of these hypothetical cationic species.

Some typical diterpenes are shown in Fig. 3.16, together with the postulated or proven pathways of biosynthesis. The cyclizations and subsequent reactions are unexceptional: abietic acid is a common constituent of tree resins, as is cembrene (pine resins). As usual, further skeletal modifications may take place, and the metabolite taxinine, which is the main toxic constituent of the yew tree, is one such example.

Little tracer work has been carried out (except in the gibberellin field, see below), but virescenol A (see Fig. 3.16) incorporated label from [1-^{13}C]-acetate to produce the anticipated labelling pattern. Often a whole sequence (probably concerted) of migrations must be invoked in order to rationalize production of grossly altered skeletal types, e.g. rosenonolactone [165] (eqn 3.9). Experiments with 5(S)- and 5(R)-monodeutero-MVA samples coupled with deuterium n.m.r. analysis, established that the 5-pro-(R)-hydrogen of MVA (corresponding to H$_a$) becomes the (Z)-hydrogen of the ethenyl group in [165]. Conversely, the 5-pro-(S)-hydrogen (corresponding to H$_b$) becomes the (E)-hydrogen. The cyclization thus proceeds with the *anti*-stereochemistry expected for an S$_N$2′ process.

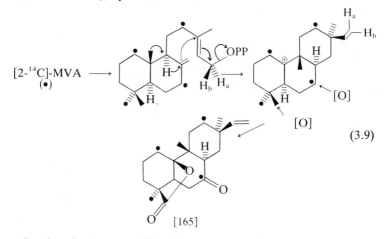

$$(3.9)$$

By far the best studied diterpenes are the *gibberellins*. These diterpenoid acids were originally isolated from the fungus *Gibberella fujikuroi*, a pathogen which causes overgrowth of rice seedlings (tall straggly plants are obtained). The gibberellins were shown to be the causative agents of this syndrome, and have since been isolated from many healthy plants as well. Indeed they seem to be ubiquitous plant hormones, responsible for increased growth, and induction of flowering, amongst other things. Their biosynthesis has been exhaustively studied,

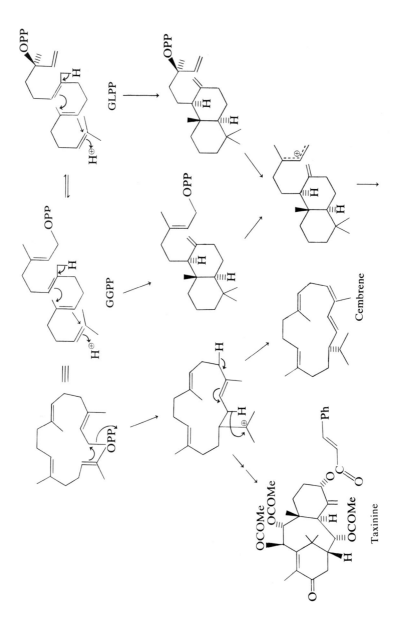

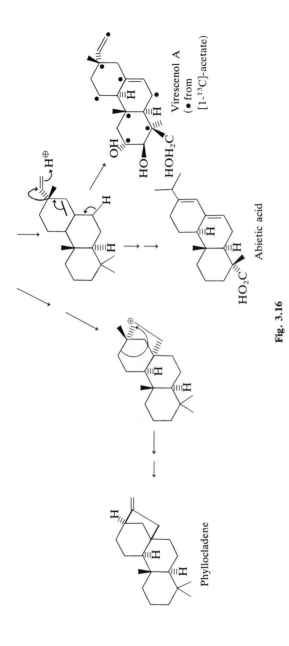

Virescenol A
(● from
[1-^{13}C]-acetate)

Abietic acid

Phyllocladene

Fig. 3.16

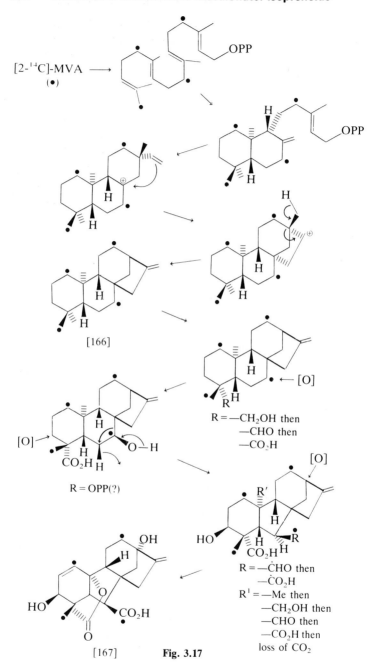

Fig. 3.17

and the pathway proceeds via kaurene [166], a metabolite with $5\beta,10\alpha$ stereochemistry. The currently accepted route which yields gibberellin A$_3$ [167] (possibly the most active of the forty or so known gibberellin hormones), is shown in Fig. 3.17.

Sesterterpenes

A relatively small number of C$_{25}$ isoprenoid metabolites occur in Nature: these are called *sesterterpenes*. They presumably arise via coupling of GGPP and IPP (C$_{20}$ + C$_5$). Few simple compounds are known, and the ophiobolins (e.g. ophiobolin B [168] of fungal origin, comprise the largest group of sesterterpenes thus far encountered. In addition, many insect waxes contain interesting sesterterpenes, e.g. gascardic acid [169], and a

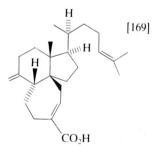

[168]

series of furanoids have been isolated from sponges of the genus *Ircinia*, e.g. variabilin [170].

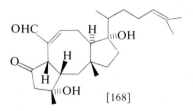

[169]

CO_2H

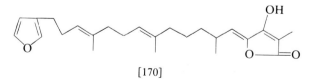

[170]

Steroids and triterpenes

Amongst this group of metabolites are compounds such as cholesterol, the mammalian sex hormones, the vitamins D, etc., which are vital for the integrity of many organisms, especially animals. These are thus not secondary metabolites in the classical sense of being compounds of secondary importance. Nonetheless, possession of the necessary metabolic pathways which produce cholesterol and the triterpenes is by no means universal: most insects lack the pathways, and must ingest cholesterol or plant sterols (phytosterols) in order to satisfy their requirement for triterpenes or triterpene metabolites.

Cholesterol and other steroids are not true triterpenes in that they possess C_{27}–C_{29} skeletons (or smaller skeletons derived from these parent species), rather than a C_{30} skeleton; but they are best considered along with the true triterpenes since all are derived from the same C_{30} precursor: squalene [171] (see Fig. 3.18). This polyolefin was for many years a curiosity, found in large quantities in shark liver oil, but in trace amounts elsewhere. It was suggested that, by cyclization and loss of three methyl groups, cholesterol might be obtained. Indeed, [1-^{14}C]- and [2-^{14}C]-acetate were incorporated into squalene, which was in turn incorporated efficiently into cholesterol by cell-free extracts of rodent liver. Subsequent work by Bloch, Lynen, Cornforth, and Popják has led to the delineation of the pathway.

Squalene is derived from two farnesyl PP units, but these must be joined in the unusual 'head-to-head' fashion. The stereochemistry of this process is known from tracer studies, and it is believed to proceed via the intermediate presqualene pyrophosphate [172]. A reasonable, though as yet speculative mechanism for the entire process is shown in Fig. 3.18. (cf. the mechanism shown in Fig. 3.7).

The polycyclic structures formed from squalene can all be rationalized in terms of the ways in which squalene may be folded (pseudo chair and boat conformations) on the enzyme surface, with due consideration given to the stereoelectronic requirements for cyclization. This is usually initiated by acid-catalysed ring opening of squalene monoepoxide [173] (Fig. 3.19), and probably occurs via a series of carbocationic intermediates. It is worth noting that experiments have shown that epoxidation can occur at either end of the squalene molecule, suggesting that it is released from the synthetase enzyme complex prior to epoxidation and cyclization. Once away from the asymmetric environment of the enzyme, the molecule may behave like any other symmetrical, organic molecule, which would explain the propensity for epoxidation at either end. However, only the (3S)-stereoisomer is utilized by a wide variety of systems. The various cyclization modes are illustrated in Fig. 3.19.

The initially formed cationic species [174] and [175] are purely

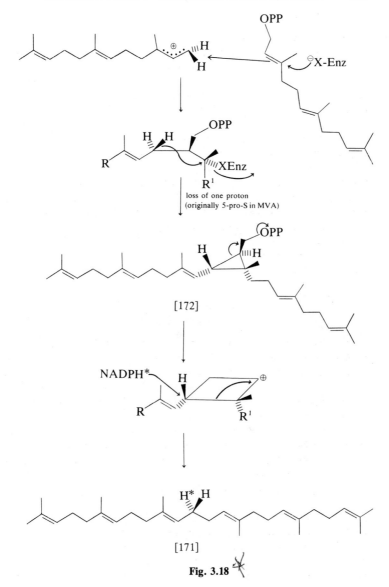

Fig. 3.18

hypothetical, and enzyme-bound nucleophiles probably participate. However, incipient electron deficiency at these sites in the molecules is supported by the fact that nucleophiles are often incorporated at equivalent positions when model substrates are cyclized *in vitro*. Numerous such

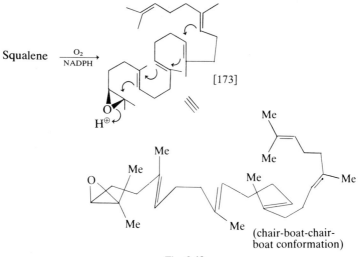

Squalene $\xrightarrow[\text{NADPH}]{\text{O}_2}$ [173]

(chair-boat-chair-
boat conformation)

Fig. 3.19

cyclizations of squalene-type compounds have been accomplished, and an example is shown in eqn (3.10). It is noteworthy that these *in vitro* processes usually yield a major product which has the same overall,

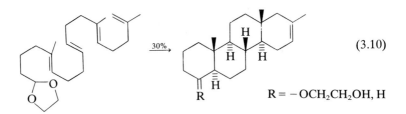

$$\xrightarrow{30\%} \qquad (3.10)$$

$$R = -OCH_2CH_2OH, H$$

relative, stereochemistry as cholesterol (i.e. *trans–anti-trans*). Any analogue of squalene which satisfies certain structural and stereo-electronic requirements should cyclize to an analogue of cholesterol, and through the use of such analogues the minimum structural requirements for steroid formation by the cyclase enzyme have been identtified [176].

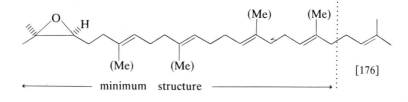

minimum   structure

[176]

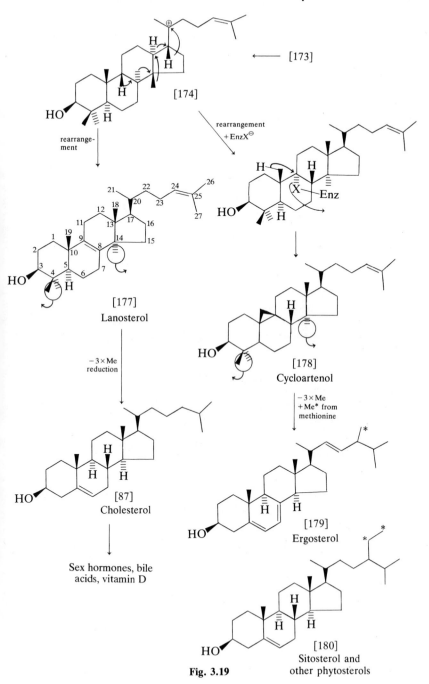

Fig. 3.19

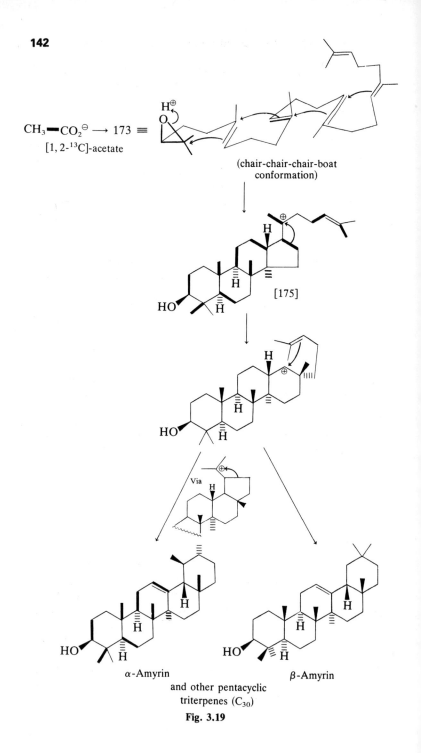

(chair-chair-chair-boat
conformation)

[175]

Via

α-Amyrin
and other pentacyclic
triterpenes (C₃₀)

β-Amyrin

Fig. 3.19

In general, animals employ the lanosterol [177] to cholesterol [87] pathway, while plants utilize the cycloartenol [178] to phytosterol (e.g. ergosterol [179] in yeasts, sitosterol [180], in plants) pathway. Fungi and certain other lower organisms appear to utilize either or both pathways, and this is probably due to the lower specificities of the enzymes they employ. Many triterpenes derive from intermediate [175], though different modes of folding are often employed, and partially cyclized triterpenes are also common. The scheme shown in Fig. 3.19 has been substantiated using [1,2-^{13}C]-acetate as precursor.

The various modes of biosynthesis are still not completely understood, and the sequence of reactions and intermediates probably differ slightly between organisms. However, a few general points may be made concerning steroid biosynthesis:

(a) The C-24:25 double bond suffers a *cis* reduction (cholesterol pathway); $H^{\oplus}$ from the medium adds at C-24, and $H^{\ominus}$ from NADPH adds at C-25.

(b) Alternatively, alkylation may occur at C-24 (phytosterol pathway), with successive methyl groups derived from S-adenosyl methionine; β-sitosterol [180] may be derived as shown in eqn (3.11).

(c) The two methyl groups that are lost from C-4 are almost certainly removed via the oxidative sequence—Me to —CH_2OH to —CHO to —CO_2H, and decarboxylation; whilst the methyl at C-14 is lost as methanoic acid (see eqn 3.12); but the order of removal varies from system to system.

(d) A series of olefinic intermediates have been postulated, and metabolites of this kind have been isolated. It is unlikely that there is one pathway, but rather a series of similar routes, one of which is shown in eqn (3.12).

We shall now consider in more detail some steroidal types and their biological functions.

Biological functions of steroids

When considering the possible biological functions of steroids (and triterpenes), we are confronted with the fact that on a total weight basis, the majority of sterols (i.e. steroid alcohols) are long-lived; they do not appear to be metabolized to any extent. It is believed that these sterols have a vital role in maintaining the structural integrity of most membraneous structures in organisms. They also appear to assist in the regulation of the permeability of these membranes to various ions.

All eucaryotes (organisms which have discrete membrane-bound organelles, e.g. a cell nucleus) synthesize sterols, or have an absolute requirement for them in their diet. Even some procaryotes (organisms

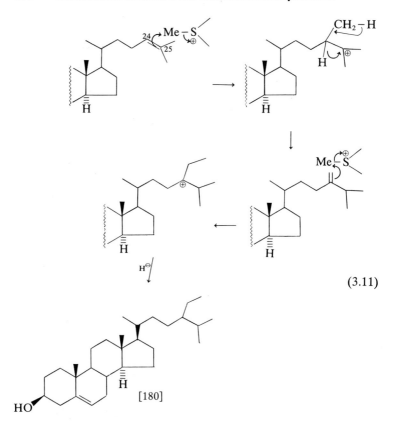

(3.11)

[180]

possessing no membrane-bound organelles) have the capacity to cyclize squalene, although they contain no sterols. Thus *Tetrahymena pyriformis* will metabolize squalene to produce tetrahymanol [181], a pentacyclic analogue of lanosterol [177], and this compound probably has a vital structure role in the organism. This phenomenon is also encountered in lower organisms, where hydroxylated carotenoids or polyprenols may serve as sterol substitutes.

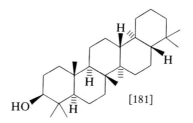

[181]

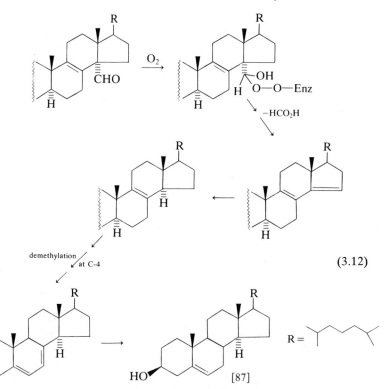

$$(3.12)$$

These long-lived sterols all possess a similar, rather flat structure, and this is depicted as [182]. Other rapidly metabolized steroids, like the sex hormones, the vitamins D, etc. are of more immediate interest here, and we shall not discuss the long-lived sterols further.

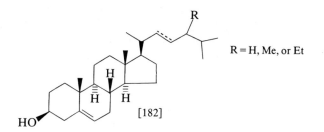

R = H, Me, or Et

[182]

In plants, algae, and fungi, a variety of phytosterols serve as substrates for further metabolism, while in animals, cholesterol is the source of all steroid metabolites.

Metabolism of sterols in plants, algae, and fungi
Several basic phytosterol skeletons are encountered and these, together
with some of the biogenetic pathways (some proven) by which they might
arise, are shown in Fig. 3.20. Phytosterols may be metabolized in two
ways: with, or without, cleavage of the side-chain.

   Metabolism without side-chain cleavage. Phytosterols are of particular dietary importance to insects and certain crustaceae which cannot

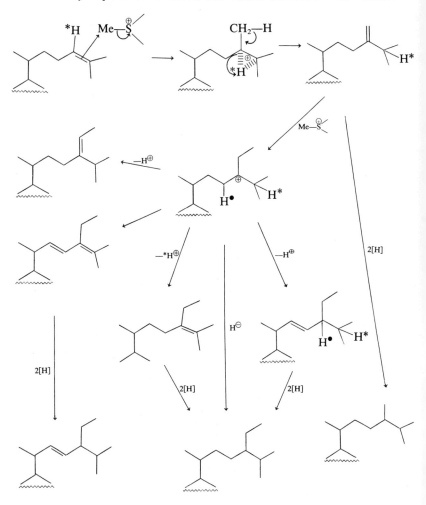

Fig. 3.20

synthesize cholesterol. These organisms can degrade C_{28} and C_{29} phytosterols to C_{27} sterols (usually cholesterol), which are then used in the synthesis of certain biologically active sterols that the organisms require. There is some evidence which suggests that the dealkylation of the phytosterols proceeds as shown in eqn (3.13).

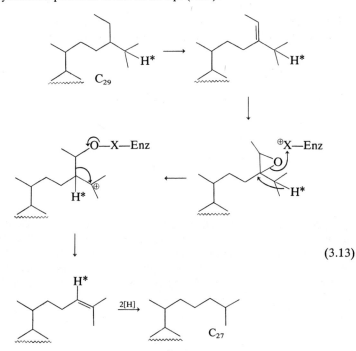

(3.13)

Some of the most interesting of these insect phytosterols are the so-called moulting hormones, or *ecdysones*, e.g. α- and β-ecdysones [183] (R = H and R = OH). These hormones, released by the insect larva, act

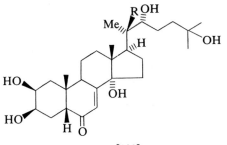

[183]

upon epidermal cells and cause initiation of moulting. The overall metabolic balance of juvenile hormone [132] and the ecdysones controls the process of larval development and metamorphosis. Many plants synthesize ecdysones, and it is not inconceivable that these metabolites confer some selective advantage on them, provided that it can be shown that feeding on these plants is detrimental to insects. Such evidence is not as yet available.

The biosynthetic pathway from cholesterol to the ecdysones is not completely understood, but probably proceeds via the intermediate [184].

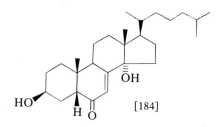

Other plant metabolites which warrant mention, and which also contain intact, though modified side-chains, are the *sapogenins*, and C_{27} *alkaloids*. The most useful sapogenin, without doubt, is diosgenin [185], which occurs as its 3-O-glycoside (sugar ether) in several types of *Dioscorea*. The metabolite is available in large amounts (the roots contain as much as 5–6 per cent by weight), and is used as starting material in a short industrial process for the synthesis of cortisone [186].

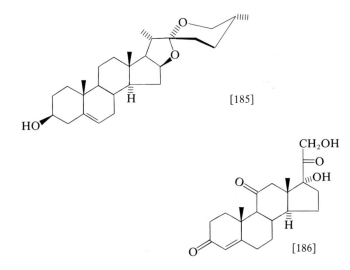

This has a variety of clinical uses, notably as an anti-inflammatory agent, but also occurs naturally in all mammals. From the discovery in 1940 of a simple route from diosgenin to progesterone (the hormone of pregnancy) [187] until 1952, when a method of introducing the 11-oxygen functionality was described, was a period of intense activity and competition amongst American pharmaceutical companies. Finally, the 'wonder drug', cortisone, was produced in useful amounts; but long-term administration to suppress inflammatory conditions like rheumatoid arthritis, caused complications. Metabolism was disturbed, and a variety of dangerous side-effects were noted. However, the technology and expertise were now available, and a whole battery of new steroidal anti-inflammatory agents has been synthesized in the ensuing quarter century. Amongst these, betamethasone valerate (Betnovate, Glaxo) [188] is one of outstanding efficacy for topical applications. Diosgenin is biosynthesized from cholesterol probably via intermediate [189], but the complete pathway has not been delineated.

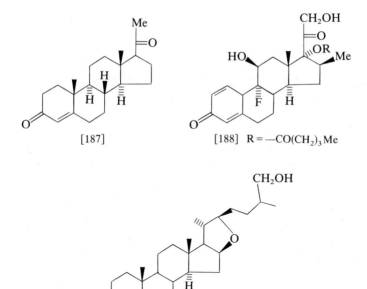

[187]

[188] R = —CO(CH$_2$)$_3$Me

[189]

The C$_{27}$ alkaloids, such as solanidine [190] (potato and tobacco plants), and tomatidine [191] (tomato) are obviously related to diosgenin in terms of structure, and are also derived from cholesterol, by routes unknown.

These metabolites usually occur as their 3-O-glycosides, e.g. solanin and tomatin, and many of them are fungitoxic. They may protect the

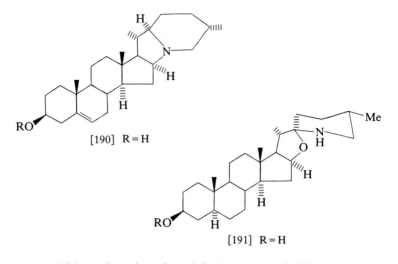

[190] R = H

[191] R = H

plants which produce them from infection, or may inhibit growth of a fungus after invasion of the plant. In addition, some are teratogenic, that is they can cause congenital malformation *in utero*. Thus, one called quite appropriately cycloposine, produces cyclopian malformations in lambs born to ewes which have fed upon plants containing this metabolite.

Metabolism with side-chain cleavage. Cholesterol [87] and β-sito-sterol [180] are metabolized in higher plants to pregnenolone [192], progesterone [187], and thence to *cardenolides, bufadienolides*, and C_{21} *alkaloids*, via pathways which are, as yet, not completely understood. Some of the most interesting compounds are shown in Fig. 3.21. The so-called cardiac glycosides, such as digitoxin and digoxin are cardenolides; and are well known for their use in medicine as cardiac stimulants. They also induce vomiting and affect the central nervous system, and it is certain that these combined effects are unpleasant (and even fatal) to grazing animals. Synthesis of cardenolides may thus confer an advantage on the plants concerned, e.g. the foxglove, *Digitalis pupurea*, in that they may be spared the attentions of browsing herbivores, which have learnt to avoid them.

In several instances animals, reptiles, and insects have been shown to be immune to the effects of these metabolites. Thus the common European toad contains bufotoxins (see Fig. 3.21), and the Monarch butterfly may contain cardenolides if its larval progenitor has consumed the foliage of plants containing these metabolites. In both cases the creatures are quite unpalatable to a wide range of predators.

A number of other very toxic metabolites of this type occur naturally, and two examples will suffice: samandarin [193], a deadly neurotoxin from the skin of the salamander, and batrachotoxin [194], from a South American frog. The latter compound is used as an arrow poison by South American Indians, and is thus of importance in human ecology as well as playing a defensive role in the frog.

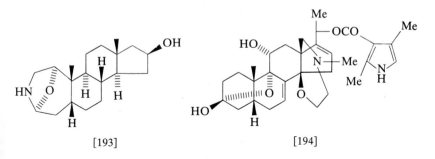

[193]   [194]

The C_{21} alkaloids, such as conessine (see Fig. 3.21), are also derived from cholesterol (and probably from sitosterol and other phytosterols), but they have no known function, apart from their probable deterrent effect on herbivoires.

Metabolism of sterols in animals
Metabolism of cholesterol in animals produces two major classes of metabolites: the steroid hormones, and the bile acids.

Steroid hormones. The sequence, cholesterol to pregnenolone, [192], is generally accepted, though the oxidative mechanism shown in Fig. 3.22 is not fully substantiated. (The enzyme cytochrome P450 mediates many steroid oxidations.) Subsequent nuclear modification gives rise to the various steroid hormones. Oestradiol and progesterone are human female sex hormones, and testosterone and androst-5-en-3β-ol-17-one are male sex hormones. They influence development at puberty, and in the case of the female hormones, regulate the menstrual cycle. Synthetic oestrogens and progestins such as mestranol [195], and norethindrone [196] are the basis of the contraceptive pill, and function by disturbing the delicate balance of natural oestrogens and progesterone during the menstrual cycle. The availability of diosgenin [185] has also proved invaluable in this area of industrial synthesis. Another route commences with cholesteryl acetate [197], and utilizes a soil microorganism to effect demethylation at C-19 (eqn 3.14) and so produce oestrone [198], and thence mestranol and norethindrone. The *in vivo* conversion of pregnenolone [192] into oestrone also proceeds via a 19-hydroxylated intermediate, with subsequent loss of C-19 as formic acid. This route to the oestrogens has been probed by the use of $^{18}O_2$. Since about one-third of mammary

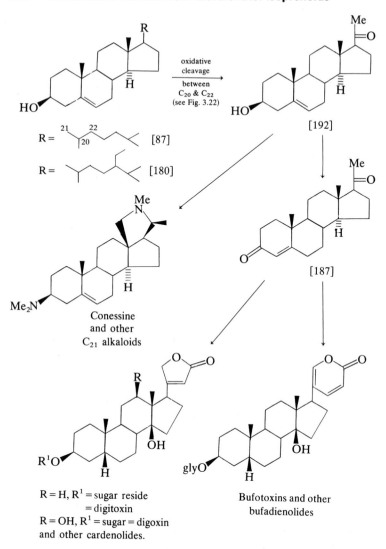

R = $\begin{smallmatrix}21 & 22\\ & 20\end{smallmatrix}$ [87]

R = [180]

[192]

[187]

oxidative cleavage between C_{20} & C_{22} (see Fig. 3.22)

Conessine and other C_{21} alkaloids

R = H, R¹ = sugar reside
= digitoxin
R = OH, R¹ = sugar = digoxin
and other cardenolides.

Bufotoxins and other bufadienolides

Fig. 3.21

tumours (i.e. breast cancers) require a supply of oestrogens for growth, one effective kind of chemotherapy relies upon the use of inhibitors of the aromatase enzymes. One such inhibitor is 4-hydroxyandrostenedione (X = OH), and this is shown in Fig. 3.22.

It is interesting to note that the original sample of oestradiol (see Fig. 3.22), 12 mg in all, was extracted from four tonnes of sow ovaries, and the

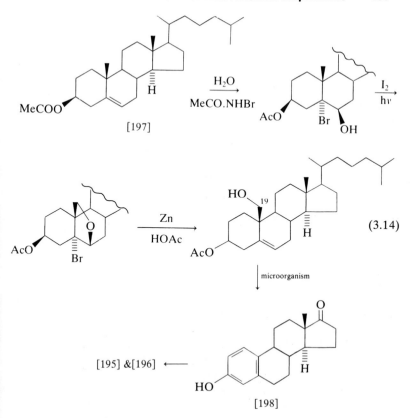

(3.14)

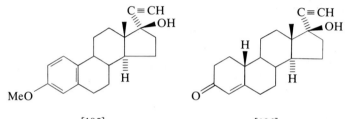

first 15 mg of androsterone came from 15 000 cm³ of male urine! A far
cry from the vast amounts of synthetic oestrogens and progestins that are
produced and consumed today.

The other two steroid hormones which appear in Fig. 3.22, cortisone,
and aldosterone, have a broader spectrum of activity than the sex
hormones. They help to control the metabolism of carbohydrates,

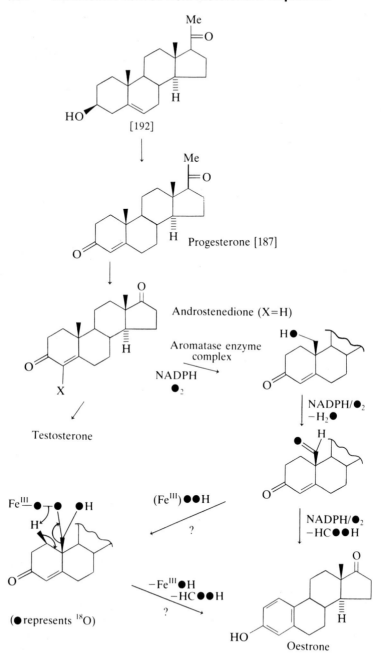

Fig. 3.22

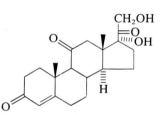

Testosterone

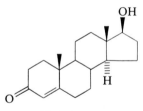

Cortisone [186]

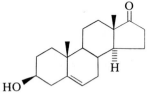

Androst-5en-3-β-ol-17-one

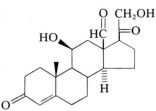

Aldosterone

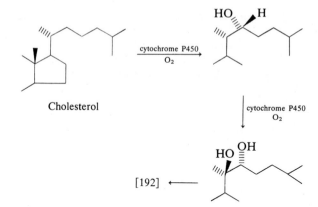

Cholesterol

[192]

Fig. 3.22 (*continued*)

proteins and fats, as well as regulating salt and water retention and excretion. They are synthesized in the adrenal glands, and are often known as the adrenocorticoids.

Although we have confined our attentions to animals, primarily Man, it is worth noting that the sex hormones are widespread in plants and microorganisms. Here they function as general anabolic agents, i.e. they stimulate *de novo* biosynthesis. They also produce a variety of odd effects including promotion of flowering in some plants.

Bile acids. The bile acids (or rather their sodium salts) are major constituents of bile, and the only ones which are active in digestion. They emulsify fats that are present in the part-digested food mass, producing tiny droplets of fat which are more susceptible to enzymatic cleavage. They are synthesized from cholesterol in the liver, and together with the other constituents of bile, are stored in the gall bladder. Little bile is excreted, but is reabsorbed and recycled through the liver.

One biosynthetic pathway is shown in Fig. 3.23: hydroxylation at C-7 is believed to be the rate-determining step. A possible alternative pathway, in which side-chain cleavage precedes complete ring hydroxylation, has also been suggested.

Vitamin D

Vitamin D is the common, collective name for a group of structurally similar sterols which are necessary for the optimum absorption and metabolism of calcium and phosphorus by animals. Deficiency leads to rickets in children, and osteomalacia in adults. Vitamin D_2 [89a] is present in plants and yeasts after UV irradiation (i.e. from sunlight), while cod liver oil is a rich source of vitamin D_3 [89b].

Vitamin D is added to many foods, and in addition, the action of sunlight on 7-dehydrocholesterol [199], which is present in the skin, produces vitamin D_3. This is in fact the method of industrial synthesis, and as mentioned in Chapter 1, the stereochemistry of this photochemical 6π ring-opening reaction, and others like it, led Woodward and Hoffmann to expound the rules for the conservation of orbital symmetry in concerted reactions. This particular process is a conrotatory cycloreversion (eqn 3.15).

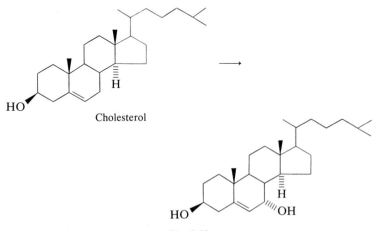

Cholesterol

Fig. 3.23

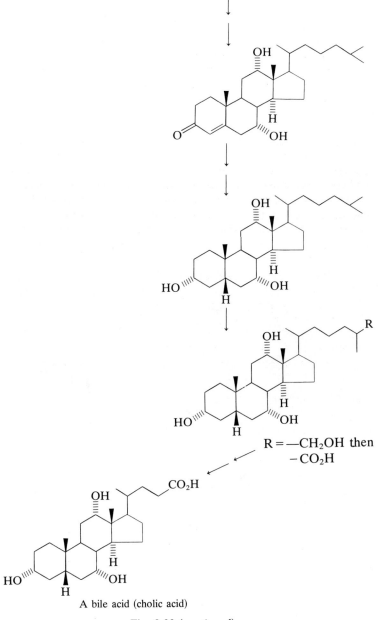

R = —CH₂OH then — CO₂H

A bile acid (cholic acid)

Fig. 3.23 (*continued*)

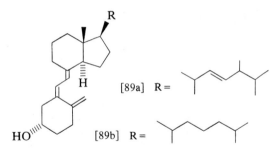

Vitamin D_3 itself is not the active species *in vivo*, but two hydroxylated metabolites (1,25-dihydroxy- and 1,24,25-trihydroxy vitamin D_3) are responsible for many of the biological activities of vitamin D_3. In addition, synthetic 1-α-hydroxy-vitamin D_3 is cheap to produce and almost as potent as these metabolites, and might have therapeutic potential. These compounds assist in the assimilation of calcium ions from the diet and stimulate transport of calcium ions *in vivo*, and in conjunction with

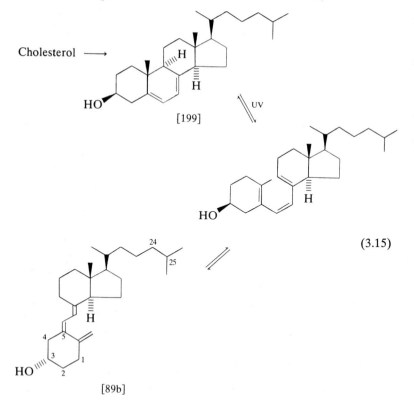

(3.15)

parathyroid hormone (a polypeptide released by the parathyroids), they mobilize calcium ions from bone.

Recently, a South American plant, *Solanum malacoxylon*, has been shown to produce a metabolite that resembles the vitamins D in its biological activity: cattle consuming the leaves of this plant exhibit all of the symptoms of hypervitaminosis D. The metabolite is a 1,25-dihydroxy-vitamin D_3-glycoside.

It is pertinent to point out that the isoprenoids, and in particular the steroids and triterpenes, with their complex three-dimensional structures, have an obvious potential as 'informational molecules'. That is, an almost infinite variety of differently shaped molecules are possible. Most other secondary metabolites contain aromatic rings, or long alkyl chains, with few if any chiral centres, and thus little potential for variation of shape. It is not surprising then, that more examples of compounds with a known biological function are found amongst the isoprenoids than in any other class of secondary metabolites. We should also not be surprised that biosynthetic pathways to the basic triterpenoid structures have existed for almost as long $(1 \times 10^9$ years) as life has existed on this planet. Triterpenoid skeletons have been found in ancient sediments, fossils, and spores, and usually they possess the same stereochemistry as their modern counterparts.

Carotenoids

The carotenoids are, with few exceptions, C_{40} polyolefinic metabolites. They lack the complex three-dimensional structures of the triterpenes and steroids, but have the potential for a vast number of geometric isomers. In fact most of them have the stable all-*trans* configuration. They occur most commonly in green plants and certain lower organisms (algae, bacteria, and fungi), and many are coloured due to their extended, conjugated systems. The most familiar carotenoid is undoubtedly β-carotene [200], the orange-red pigment of carrots, which occurs in most green plants. The other very common bicyclic carotene is α-carotene [201].

The carotenes are almost certainly essential, accessory pigments in photosynthesis, but their exact role is not known. Similarly, they may be involved in other processes which utilize sunlight, and may help to prevent cell damage due to photo-oxidation.

Biosynthesis proceeds via condensation of two molecules of geranyl geranyl PP (C_{20}) [160] to produce phytoene [202]. Both 15,15'-*Z* and 15,15'-*E* isomers appear to be formed *in vivo*, with concomitant loss of one 1-pro-(S)-hydrogen (originally 5-pro-(S) of MVA) from each molecule of GGPP (15,15-*Z*-phytoene), or of one 1-pro-(S)-hydrogen and one 1-pro-(R)-hydrogen (15,15'-*E*-phytoene). Various schemes have been proposed to explain these stereochemical findings, including one

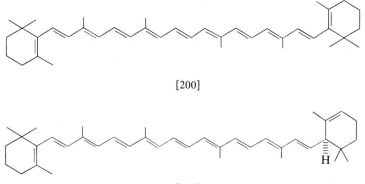

[200]

[201]

pathway via prephytoene pyrophosphate [203], a homologue of presqualene pyrophosphate [172] (see Fig. 3.18). Mutant strains of *Neurospora crassa*, which cannot produce phytoene, accumulate this compound, and the pathway shown is the most reasonable one. When radio-labelled prephytoene was fed to a mycobacterium, labelled lycopersene (Fig. 3.24) was the final product; but under natural conditions, phytoene is almost certainly the primary product, formed without the intermediacy of lycopersene. It is interesting to note that lycopersene (the homologue of squalene) was also produced when GGPP was fed to squalene synthetase from bakers' yeast: the natural substrate, farnesyl PP, leads of course to production of squalene.

Once formed, phytoene is desaturated (as shown in Fig. 3.24) to produce lycopene via stereospecific removal of pairs of hydrogens, alternatively to the left and right of the central triene system. These hydrogens were originally the 5-pro-(R)-and 2-pro-(S)-hydrogens of MVA. Cyclization of phytoene to produce polycyclic systems (cf. cyclization of squalene to sterols and triterpenes) does not occur, presumably because the rather rigid geometry of the central triene precludes adoption of favourable conformations.

Lycopene, a major pigment of tomatoes, is the final acylic carotenoid, and subsequent cyclization produces α- and β-carotenes. Two possible modes of ring closure are shown in eqn (3.16). However, the stereochemical details of the various cyclization reactions are still poorly understood; and although these various processes appear to be similar, the stereochemistry is probably different in each case. This is surely the result of a number of factors, including the initial modes of folding, and nature of the species which initiates cyclization. It is also possible that cyclization may occur at an early stage, at least in some organisms, and pathways from neurosporene which bypass lycopene are known. Monocyclic carotenes also exist.

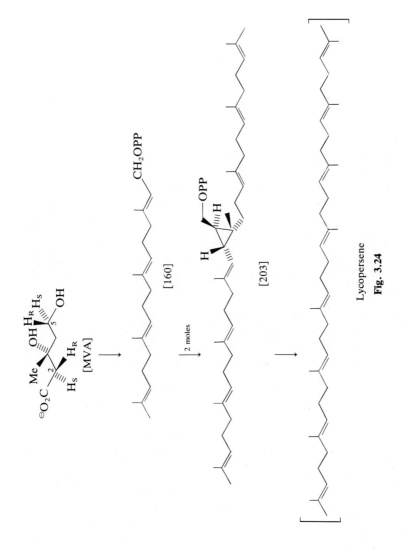

Lycopersene

Fig. 3.24

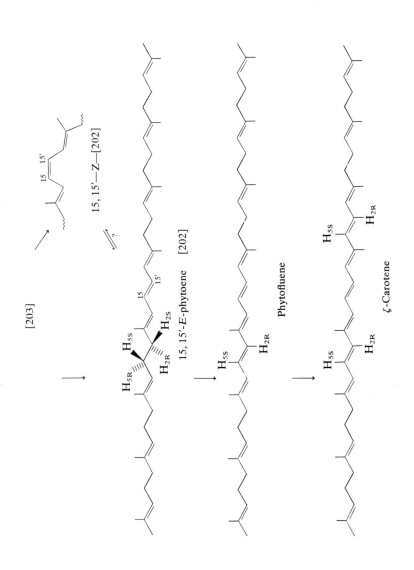

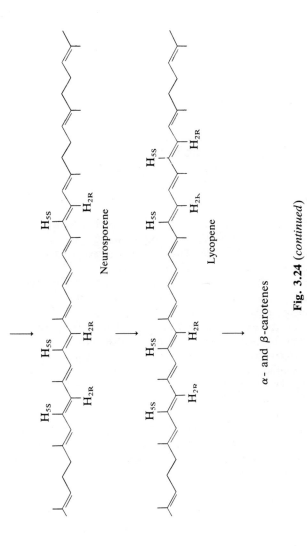

Neurosporene

Lycopene

α- and β-carotenes

Fig. 3.24 (*continued*)

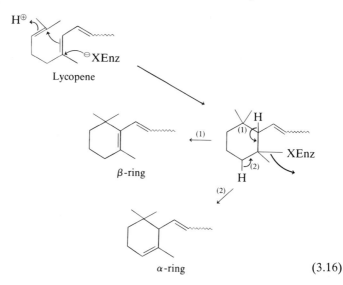

(3.16)

Further modification of carotenes is common, and the oxygenated derivatives (possibly the products of photoprotective processes) are known collectively as *xanthophylls*. The process is as expected aerobic, and pathways from α- and β-carotenes are shown in eqn (3.17) Such parallel oxidative transformations provide a good example of a 'metabolic grid': a few, rather non-specific enzymes catalyse an identical series of transformations of similar substrates.

More drastic oxidative modification leads to metabolites like capsanthin [204] from red peppers. It is conceivable that this derives via rearrangement of an epoxide intermediate as shown in (eqn 3.18).

The allenic xanthophyll fucoxanthin [205], occurs in brown algae and other lower organisms, and is one of the most abundant natural carotenoids. It is believed to have an essential role as an accessory pigment in algal photosynthesis.

Most of these modified metabolites still retain a C_{40} skeleton, but there are a number of metabolites which are produced by degradation of carotenoids, and these usually possess a C_{20} skeleton. Examples include vitamin A and the trisporic acids.

Vitamin A
Animals require vitamin A [88] or a carotenoid precursor for normal growth and vision. An early sympton of vitamin deficiency is a decreased ability to see in dim light. This is not surprising since vitamin A aldehyde, or retinal [206] has an essential role in vision (see Fig. 3.24). The

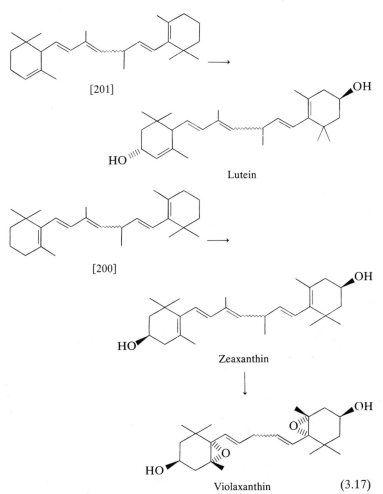

[201]

Lutein

[200]

Zeaxanthin

Violaxanthin (3.17)

11-*cis* isomer combines with a protein, opsin, via a Schiff-base linkage, to form rhodopsin. Rhodopsin is located in the rods and cones of the retina, and when exposed to light, geometric isomerism occurs, producing all-*trans*-retinal with concomitant release of the protein moiety. This conformational change is somehow translated into a nerve impulse, probably due to subtle changes in the level of hydrogen ions, and transmitted to the brain. 11-*cis*-Retinal is regenerated in a further sequence of reactions (Fig. 3.25), and the net result is vision. The actual kinetics and stoichiometry of the 'proton pumping' are not known, but the geometrical isomerization is believed to cause a change in charge

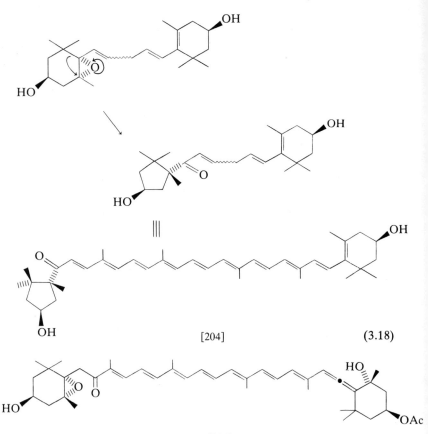

[204]   (3.18)

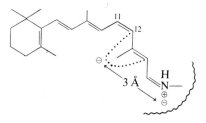

[205]

separation, which results in changes in the pk_a values of various amino acid side-chains and consequent movement of protons. A model for the *cis*-form is shown below:

Vitamin A is almost certainly formed by symmetrical oxidative cleavage of β-carotene (at the central double bond).

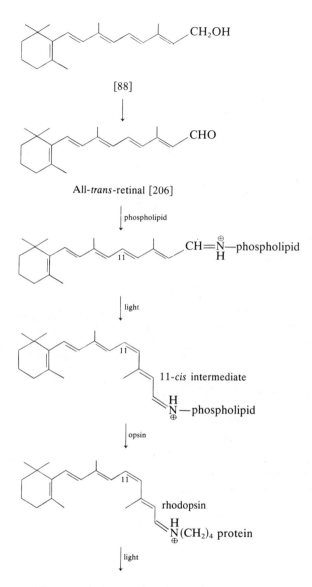

Fig. 3.25

Trisporic acids
These metabolites e.g. triporic acid [207], are the principal sex hormones of the fungi Mucorales. Fusion of two cell types (+ and −) is involved in sexual reproduction, and the trisporic acids are implicated in regulation of the production and association of the two mating types. They are formed from β-carotene via retinal [206] with further loss of two carbon atoms, and since they also stimulate carotenoid biosynthesis, their biosynthesis is self-amplifying.

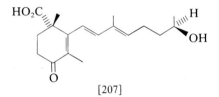

[207]

Abscisic acid
Abscisic acid [208] is a plant hormone of almost universal occurrence. It has a role in regulation of plant growth since it induces and regulates dormancy of buds, and also inhibits germination, in many species. Labelling studies, including the use of [1,2-^{13}C]-acetate, have shown that it is probably produced directly from MVA, but formation from a carotenoid precursor was formerly considered since compound [209], which probably derives from violaxanthin (see eqn 3.17), is converted into abscisic acid in tomato and wheat. The epoxide oxygen atom becomes the tertiary hydroxyl oxygen atom of abscisic acid.

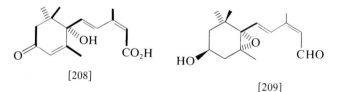

[208]

[209]

Certain other important natural products, such as vitamin E, vitamin K, cannabinoids (from cannabis), and ergot alkaloids contain C_5 units derived from MVA. These will be considered in Chapter 6 since they are products of 'mixed metabolism': that is, their skeletons comprise structural units derived via two or more separate metabolic pathways.

Problems

3.1. Feeding experiments with [2-^{13}C]-MVA (*) and [2-^{13}C]-acetate (•) have established the labelling pattern, shown below, for helicobasidin, ex. *Helicobasidium mompa*. Suggest a reasonable biosynthetic pathway.

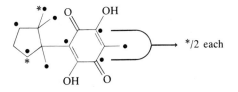

[Jankowski, W. C., Suzuki, K. T., and Tanabe, M. (1973). *Tetrahedron Lett.*, 4723]

3.2. The antifungal lactone shown below could conceivably be a sesquiterpene (with one additional carbon atom) or a degraded diterpene. Use the labelling pattern and other data in order to decide which alternative is the most likely. Propose a biosynthetic pathway. (4 moles of [2-^{14}C,5-^{3}H$_2$]-MVA are incorporated, four tritium atoms are lost.)

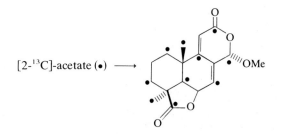

[Kakisawa, H., Hayashi, T., Ruo, T., and Sato, M. (1973). *Chem. Commun.*, 802]

3.3. The wood rot fungus *Fomes annosus* is responsible for much damage to pine forests in the south-eastern United States. It produces fomannosin, which incorporated label from [1,2-^{13}C]-acetate. How is this metabolite biosynthesized? (Hint: (•) denotes a carbon atom which derives from C-2 of MVA.)

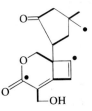

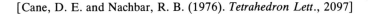

[Cane, D. E. and Nachbar, R. B. (1976). *Tetrahedron Lett.*, 2097]

3.4. [4,5-^{13}C)]-MVA was incorporated into cyclonerodiol by the fungus *Fusarium culmorum*, and the labelling pattern shown below was produced. Suggest a biosynthetic route to this metabolite.

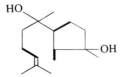

[Evans, R., Hanson, J. R., and Nyfeler, R. (1976). *Perkin Trans. I*, 1214.]

3.5. Aphidicolin is a diterpenoid from the fungus *Cephalosporium aphidicola*, and possesses both antiviral and antitumour activities. When [1,2-^{13}C]-acetate was employed as substrate the labelling pattern shown was obtained. Use this data to define the constituent isoprene units, and propose a possible biogenetic pathway.

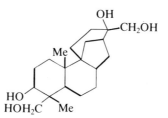

[Ackland, M. J., Hanson, J. R., and Ratcliffe, A. H. (1984). *Perkin Trans. I*, 2751]

3.6. The fungus *Marasmius alliaceus* produces the antitumour agent alliacolide. Identify the possible isoprene units, and suggest how you might establish which arrangement is more likely.

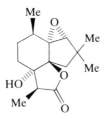

[Bradshaw, A. P. W., Hanson, J. R., and Sadler, I. H. (1981). *Chem. Commun.*, 631]

3.7. Fusicoccin is produced by the fungus *Fusicoccum amydali*, which causes plant-wilting. Incorporation experiments with [2-¹³C]-acetate produced enhanced signals for those carbon atoms shown. Propose a biogenetic pathway to this metabolite.

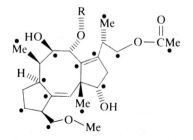

[Barrow, K. D., Jones, R. B., Pemberton, P. W., and Phillips, L. (1975). *Perkin Trans. I*, 1405]

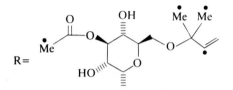

4. Metabolites derived from shikimic acid

Metabolism via the shikimate pathway gives rise to a large number of aromatic compounds related to the aromatic amino acids, phenylalanine [210], and tyrosine [211]. Many of these compounds are polyphenols, and usually possess a characteristic substitution pattern: *p*-hydroxy-, *o*-dihydroxy-, or 1,2,3-trihydroxy-. This should be compared with the typical meta-substitution pattern of polyphenols derived from acetate.

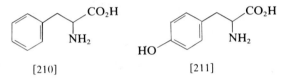

[210]   [211]

The shikimate pathway is most important in higher plants (for the production of lignins), a fact which may be understood in terms of the availability of the starting materials for the biosynthesis of shikimic acid, i.e. erythrose-4-phosphate [212], and phosphoenolpyruvate [213]. These are both involved in the primary metabolism of sugars, and have key roles in the carbon assimilation cycle of photosynthesis, a process confined mainly to higher plants and algae.

Most of the intermediate steps on the pathway to shikimic acid have now been delineated, and a reasonable biosynthetic route is shown in Fig. 4.1, though the chemistry is undoubtedly more complex. Shikimic acid itself [214], is found in Nature; and quinic acid [215], an end-product of metabolism, is commonly found (as the alcohol moiety) in ester combination with acids formed from shikimate, though it also occurs in the free form.

The main pathways of shikimate metabolism are shown in Fig. 4.2. After initial vinyl ether formation of shikimic acid 3-phosphate with phosphoenol pyruvate, and subsequent elimination of phosphate (this is a *trans*-1,4-elimination of a proton and phosphate, and is probably a two-step process), an intramolecular rearrangement of chorismate [216], produces prephenate [217]. This [3,3] sigmatropic rearrangement is not unrelated to the Claisen rearrangement, and is a biological example of a symmetry allowed (Woodward–Hoffmann) pericyclic process.

The proposed mechanism for this process is shown in eqn (4.1), and a chair-like transition state has been implicated. Knowles and Berchtold, and their coworkers, prepared samples of *Z*- and *E*-tritiated chorismates

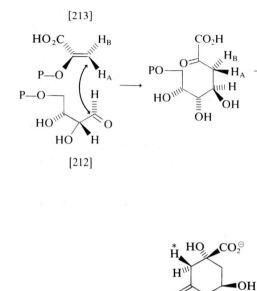

[213]

[212]

−2[H]

−Pi
+2[H]

dehydro-quinate

−H₂O
syn-elimination

dehydro-
shikimate

[215]

NADPH

[214]

Fig. 4.1

(Z-form shown in the equation), and demonstrated that the Z-hydrogen became the pro-(S)-hydrogen of prephenate and was retained through the sequence shown, while the E-hydrogen became the pro-(R)-hydrogen of prephenate and was lost. A boat-like transition state would yield complementary results. Alternatively, reduction of chorismate, and incorporation of ammonia (from the amino acid glutamine) leads, via anthranilic acid, to the essential amino acid tryptophan, which is also a precursor of the indole alkaloids (Chapter 5). This process is formally an elimination and substitution sequence. Decarboxylation of prephenate [217] is followed by aromatization, and reductive amination (i.e. introduction of ammonia under reducing conditions), to produce the amino acid phenylalanine [210]. An alternative route provides tyrosine [211], and involves reductive amination of prephenate prior to decarboxylation and aromatization. In some species, arogenate also gives rise to phenylalanine via dehydration. One enzyme, an amino acid ammonia lyase, is believed to catalyse the next reaction: the stereospecific, anti, elimination of ammonia from either phenyalanine or

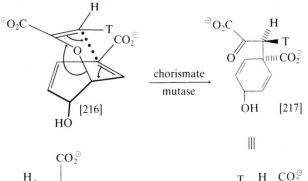

(4.1)

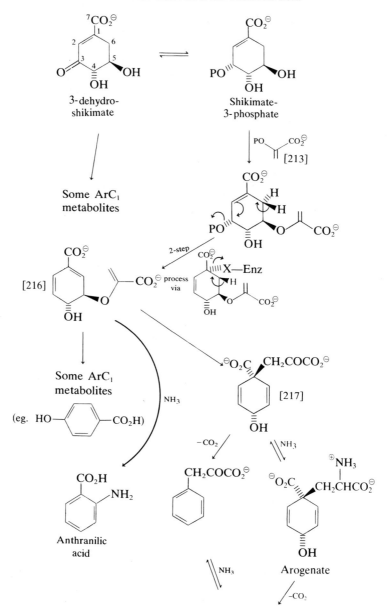

Fig. 4.2

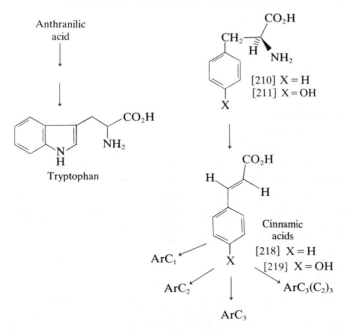

Fig. 4.2 (*continued*)

tyrosine, to yield cinnamic acid [218], or *p*-coumaric acid [219], respectively. Regulation of this enzyme phenylalanine ammonia lyase (PAL), is a critical factor in the production of shikimate metabolites, since the primary role of phenylalanine is in the production of proteins. Most organisms appear to convert cinnamic acid into *p*-coumaric acid, rather than utilizing tyrosine as a starting material. These cinnamic acids (ArC$_3$ moieties) are the precursors of most of the so-called shikimate metabolites, that is, compounds of composition ArC$_3$, ArC$_1$, ArC$_2$, and ArC$_3$(C$_2$)$_n$.

The enzymology of the pathways shown in Figs 4.1 and 4.2 has now been extensively studied, and a few salient features are worth noting. Several of the enzymes are present as complexes, such that the enzyme activities reside on a single polypeptide chain, and this is then presumably folded into discrete domains which contain the various catalytic sites. In addition, many of the key enzymes appear to exist as mixtures of isoenzymes, and possess differing sensitivities to inhibition. So, for example, chorismate mutase (eqn 4.1) has three known isoenzymes, and these vary according to whether they are inhibited by their products, the aromatic amino acids (two isoenzymes) and the cinnamic acids (one isoenzyme). Interestingly, most primitive plants only have two isoenzymes while higher plants possess all three, and the latter thus have a more effective means of controlling aromatic acid production. Although, strictly speaking, many classes of alkaloids are shikimate metabolites since they derive from phenylalanine, tyrosine, or tryptophan, we shall consider them separately (Chapter 5).

ArC₃ metabolites

Compounds of composition ArC$_3$, often known collectively as *phenylpropanoids*, are probably the most common shikimate metabolites. Almost all oxidation levels of the three-carbon side-chain are known, but the majority of the metabolites are acids or their derivatives. Some of the more important transformations of cinnamic acid [218], and *p*-coumaric acid [219], are shown in Fig. 4.3. Most of the reactions shown have been observed, and the enzymes have been isolated and characterized. The pathway from *p*-coumaric acid to aesculin, [220], has been demonstrated only recently: the previously accepted route from caffeic acid is no longer tenable.

Free phenols are rarely encountered in Nature, and it is likely that most shikimate metabolites exist as glucose ethers or in ester combination with quinic acid. It is worth noting that chlorogenic acid (caffeyl quinic acid) accounts for 13 per cent (by weight) of the soluble constituents of coffee. It is also important to realize that the metabolites isolated from natural sources are not necessarily the metabolites that are present in the living tissues. The very processes of extraction and purification must disturb the *status quo* of the organisms, and chemical changes brought about by exposure to oxygen, solvents, and changes of pH, are particularly common with phenolic metabolites. In addition, different metabolites may be produced in response to microbial infection, so the spectrum of metabolites is often characteristic of the state of health of the organism.

Of the various types of metabolite shown in Fig. 4.3, the coumarins, e.g. the archetypal coumarin [221], are of particular interest. They are widespread in Nature, and have been credited with a variety of biological activities. Thus scopoletin [222] is a potent germination stimulant (active in concentrations of 2–20 ppm), and coumarin itself undergoes metabolic changes in fermenting hay, which results in the formation of dicoumarol [223]. This has marked anticoagulant properties, and does on occasion

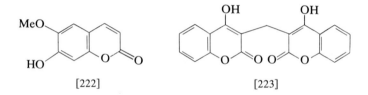

[222]   [223]

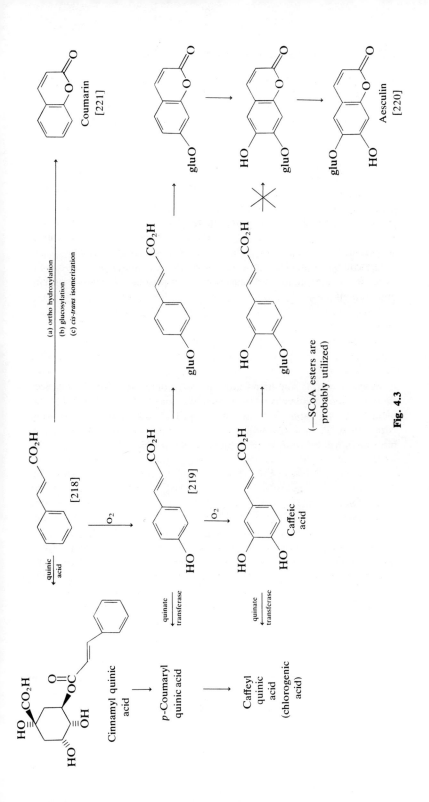

Fig. 4.3

cause the death (from internal haemorrhage) of livestock that have consumed contaminated hay. The powerful rodenticide, warfarin [224] was developed with these anticoagulant properties in mind, and has

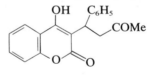

[224]

proved very effective in containing the rat population. Recently however, a strain of 'super rats' has emerged, and these appear to have developed an enzyme system that allows them to cope with this compound.

The coumarin umbelliferone [255], is also of commercial importance, since it is the major UV-adsorbing component of many sun-tan preparations.

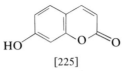

[225]

Many coumarins contain a furan ring fused onto the aromatic ring, and these are known collectively as *furanocoumarins*, e.g. psoralen [102]. As mentioned in Chapter 3, these are derived via alkylation of coumarins with DMAPP, with subsequent loss of three carbon atoms. Their biogenesis will be considered in more detail in Chapter 6.

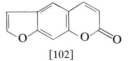

[102]

Finally, we should note that coumarins are not always derived from shikimate, as shown by the recent demonstration that metabolite [226], from the fungus *Aspergillus variecolor*, is derived from acetate (eqn 4.2).

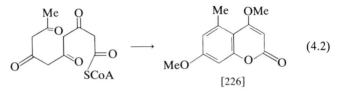

(4.2)

[226]

The isocoumarin, 6-methoxymellein [227], produced by carrots in response to stress or microbial infection, is also derived from acetate, though most normal carrot phenols are derived from shikimate.

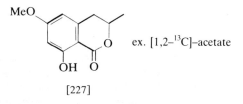

[227]   ex. $[1,2-^{13}C]$–acetate

Reduction of the side-chain of cinnamic acids
Some typical compounds formed by reduction of cinnamic acids are
coniferyl alcohol [228], eugenol [229], safrole [230], and anethole [231].
Cinnamyl-SCoA esters are reduced to the aldehydes and thence to
alcohols, and the enzymes catalysing these steps have been characterized.

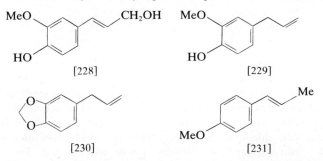

[228]   [229]

[230]   [231]

Since allyl and propenyl isomers often co-occur in the same plant, a
seemingly plausible biosynthetic pathway could be that depicted in
eqn (4.3). Some recent biosynthetic studies using *Ocimum basilicum* and
Pimpinella anisum appear to support this scheme, but previous labelling
experiments have shown that eugenol [229] is derived from a cinnamic
acid (probably ferulic-SCoA,3-methoxy,4-hydroxy-cinnamyl- SCoA) via
decarboxylation and subsequent (?) incorporation of a C_1 unit from
S-adenosylmethionine. Confirmatory evidence is awaited with some
interest. The two classes of metabolite are almost certainly formed via

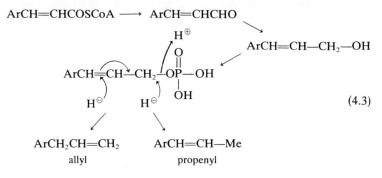

(4.3)

allyl   propenyl

different, as yet unknown, pathways, and it remains to be seen whether a common intermediate is involved.

It is interesting to note that plants of the parsley and citrus families contain a number of volatile, odiferous compounds, which are attractive to the larvae of the black swallowtail butterfly. Thus parsnip and orange contain anethole [231] and anisaldehyde [232] (an ArC_1 metabolite), both of which are attractive. The caterpillars will even attempt to feed upon blotting paper impregnated with these compounds! This was an interesting finding chemotaxonomically, since the parsley and citrus families are not considered to be closely related.

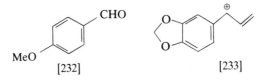

MeO   [232]   [233]

Of more commerical interest are safrole [230], a major aromatic constituent of saffron, used since biblical times as a herb, and eugenol [229], a common ingredient in oral preparations such as toothpaste and mouthwash. The metabolism of these two compounds is quite different. Eugenol has a free phenolic hydroxyl and is removed from the body after conjugation with, for example, glucuronic acid. Safrole does not possess a free hydroxyl, and is initially converted, in the liver, into 1'-hydroxysafrole (hydroxylation at the benzylic carbon centre). This is a potent carcinogen, presumably because it can give rise to a benzylic/allylic carbocation [233], which can act as an alkylating agent. For this reason, the use of oil of sassafras (high in safrole) by the food industry is now discouraged. This provides a cogent example of how two structurally similar compounds can have drastically different pathways of metabolism.

Before leaving the metabolites of composition ArC_3, we should mention two major classes of metabolites: the flavonoids, and lignins and lignans. The former are derived from p-coumaryl-SCoA and malonyl SCoA, $ArC_3(C_2)_3$, and as products of mixed metabolism, will be discussed in Chapter 6.

Lignins and lignans
In woody plants a large proportion of all ArC_3 residues are incorporated in lignins. Indeed, lignins account for a large proportion of all aromatic rings in the biosphere. They are polymeric materials held within a matrix of cellulose microfibrils, and appear to be utilized in strengthening the cell wall of the plant against external physical and chemical stresses. The three most important monomeric species utilized in the polymerization process are p-hydroxycinnamyl alcohol [234], coniferyl alcohol [228], and sinapyl alcohol [235]. Their mode of biosynthesis is shown in Fig. 4.4, and a

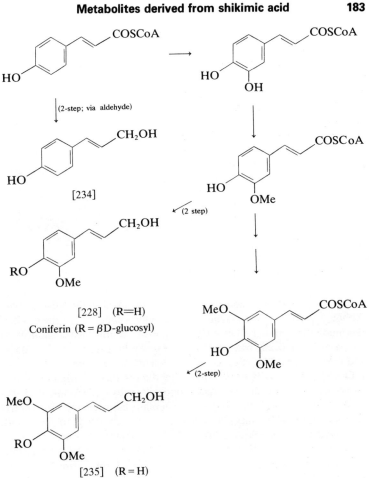

Fig. 4.4

careful study of the reduction using specifically labelled cofactor has shown that the first step involves transfer of H_S from NADPH (see eqn 1.8), while the second stage involves transfer of H_R. The compounds accumulate as their β-D-glucosides, and these are later hydrolysed, and the monomeric species oxidatively polymerized. Both C—C and C—O linkages are formed, and a representative, albeit idealized, example is shown as structure [236]. At best, a 'statistically most likely' structure is all that can be drawn, because it is impossible to elucidate the structure of lignins in their native state. The mode of oxidative polymerization is not known, but it is probably a free radical or ionic process rather than a

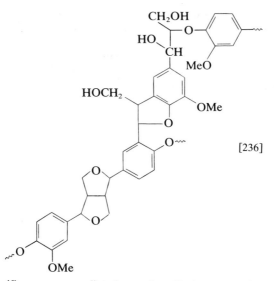

[236]

stereospecific, enzyme-mediated one since lignins are optically inactive. Although peroxidases are involved, the products appear to be the result of thermodynamic rather than enzymic control. Interestingly, lignification increases dramatically when plants come under fungal or bacterial attack, and this is probably part of an overall defence mechanism.

On the contrary, lignans, which are dimers formed from the same monomeric species, are invariably optically active. They are probably formed by stereospecific, reductive coupling between the two central carbon atoms of cinnamyl side-chains, as depicted in eqn (4.4). Representative examples are pinoresinol [237], enterolactone [238], and podophyllotoxin [239]. Enterolactone is found in human plasma, bile, and urine at levels that are close to those of the steroid hormone metabolites; in addition, there are cyclical changes in these levels in females, with peaks during the second half of the menstrual cycle and during pregnancy. There is at present no explanation for these findings, but since the compound is racemic it is believed to be derived from lignin in the diet. Podophyllotoxin is a potent antineoplastic (i.e. tumour-destructive) agent, and exerts its action by inhibiting microtubule

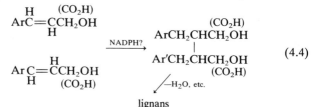

(4.4)

lignans

assembly during mitosis (cell division). Two semi-synthetic analogues are in clinical use, and are particularly effective for the treatment of testicular tumours and small cell carcinoma of the lung. These compounds and their probable biosynthetic precursors are shown in Fig. 4.5 and it has been established that two moles of ferulic acid are incorporated into [239], rather than more highly substituted precursors.

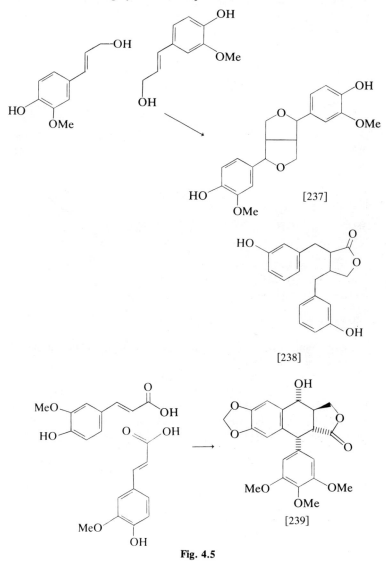

Fig. 4.5

ArC$_1$ and ArC$_2$ metabolites

Simple phenols (ArC$_0$) are relatively rare in Nature, but in the one instance where tracer studies were carried out, the compound was shown to be derived via the shikimate pathway. Thus, arbutin [240], which is the β-D-glucoside of hydroquinone, incorporated label from [^{14}C]-phenylalanine, cinnamic acid, tyrosine, and shikimic acid. Presumably, p-hydroxybenzoic acid is first formed, and subsequent oxidative decarboxylation leads to hydroquinone and thence arbutin (eqn 4.5).

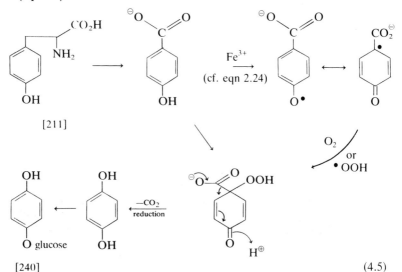

[211]

[240]   (4.5)

A large number of ArC$_1$ compounds occur in higher plants, usually in the form of esters or glycosides. They appear to be derived via degradation of the side-chain of the appropriate cinnamic acid, from dehydroshikimic acid, or from chorismic acid (Fig. 4.2). The first pathway is probably more widespread, and is shown in Fig. 4.6. Some representative examples of ArC$_1$ metabolites are: vanillin [241], from the vanilla bean, gentisic acid [242], and helicin [243] and salicin [244], both

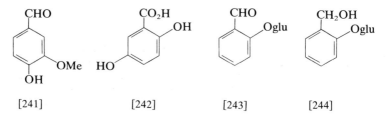

[241]   [242]   [243]   [244]

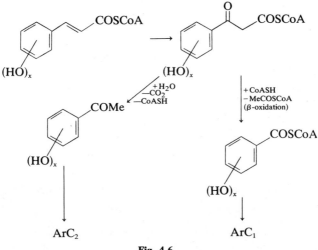

Fig. 4.6

found in species of Salicaceae (willow, poplar, etc.). It is interesting to note that extracts of willow have been used for centuries in a variety of healing preparations, and the demonstration that salicin was at least in part responsible for these effects, led to the synthesis of a more potent analogue, acetylsalicylic acid [245]: aspirin. Salicylic acid [246] occurs widely, and derives from chorismic acid (eqn 4.6), in marked contrast to 6-methylsalicylic acid [41], which is biosynthesized from acetate and malonate (Chapter 2). A particularly cogent demonstration of this was obtained using *Mycobacterium fortuitum*, which produces both compounds. After administration of [2-^{14}C]-acetate, [41] was 18 times more active than [246]; but when [^{14}C]-shikimic acid was administered, the specific radioactivity of [246] was twice that of [41]. One may merely speculate as to why Nature has chosen to synthesize two rather similar compounds via widely differing biosynthetic pathways.

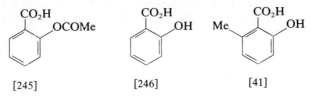

Compounds of composition ArC$_2$, which derive from shikimate, are not common, though a large number of alkaloids incorporate phenylethylamine units (ArCH$_2$CH$_2$N—), which are derived from phenylalanine or tyrosine. Labelling experiments have established that pungenoside [247], from spruce, is derived from phenylalanine; and

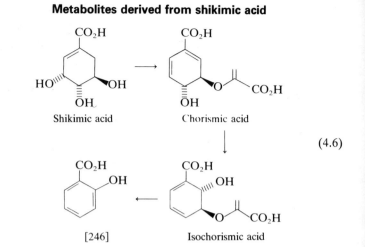

Shikimic acid   Chorismic acid

(4.6)

[246]   Isochorismic acid

acetophenone [248] is almost certainly a shikimate metabolite. The great majority of simple ArC₂ metabolites are, like 6-methylsalicylic acid, derived from acetate and malonate.

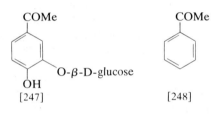

[247]   [248]

Many shikimate metabolites have a rather interesting ecological functions: they are *allelopathics*, that is chemical compounds produced by plants which 'leak' into the environment and suppress the growth or germination of other plants. These phenols are leached (by rain) from the foliage of the tres or shrubs which produce them, and make the underlying and surrounding soil barren. Thus salicylic acid [246] is present in the soil beneath oak trees, and suppresses undergrowth. Similarly, arbutin [240], *p*-hydroxybenzoic acid, vanillic acid [249], ferulic acid [250], and

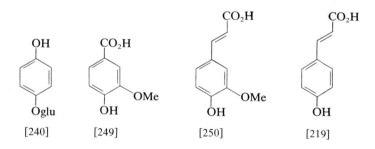

[240]   [249]   [250]   [219]

p-coumaric acid [219], are all broad spectrum allelopathics: that is they will inhibit the growth or germination of a wide range of plant species. This 'chemical warfare' between plants will be discussed in greater detail in Chapter 7.

Finally, one odd shikimate metabolite should be mentioned: chloramphenicol [251]. This is produced by a number of strains of *Streptomyces*, and is the antibiotic of choice for the treatment of typhoid.

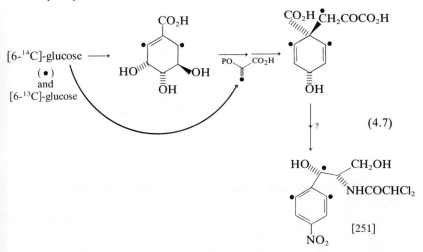

$$(4.7)$$

[251]

The results of labelling experiments are consistent with a pathway from shikimate via prephenate, but the subsequent steps of the sequence remain to be elucidated (eqn 4.7).

Problems

4.1. Chlorflavonin is a metabolite of the fungus *Aspergillus candidus*. Tracer from [3-^{14}C]-phenylalanine is incorporated (specifically) as shown. Suggest a biosynthetic route. (Hint: one ring is not derived from shikimate.)

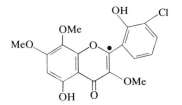

[Marchelli, R. and Vining, L. C. (1973). *Chem. Commun.*, 555]

4.2. How might you decide whether eugenin is a metabolite derived from shikimic acid or from acetate?

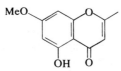

[Stoessl, A. and Stothers, J. B. (1978). *Canad. J. Bot.* **56**, 2589.]

4.3. Labelled phenylalanine was incorporated into psilotin by *Psilotum nudum*. Propose a biogenetic pathway.

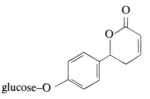

[Leete, E., Muir, A., and Towners, G. H. N. (1982). *Tetrahedron Lett.*, 2635]

5. The secondary metabolism of amino acids

The secondary metabolism of amino acids is notable for the diversity of pathways that are utilized (Fig. 5.1). Many different types of metabolites are produced, principally in higher plants, and the alkaloids comprise the largest and most exotic class. The less common pathways and metabolites occur chiefly in the lower organisms (bacteria and fungi), in which alkaloid pathways are usually absent. Those compounds, which are derived from aromatic amino acids, are formally shikimate metabolites; but, all of the metabolites considered in this chapter contain at least one nitrogen atom, an element that is usually absent from the shikimate metabolites discussed in Chapter 4.

In passing, we should note that the great majority of amino acid molecules synthesized by all organisms are utilized in the biosynthesis of essential proteins, and it seems reasonable to enquire why organisms have acquired (through evolution) the secondary pathways of metabolism. It has been suggested that the metabolites may act as a reservoir of amino

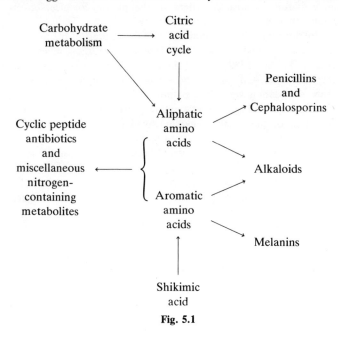

Fig. 5.1

acids for use when times are hard, but this means that the metabolites must be in equilibrium with the amino acids from which they derive. This is true in a few instances but not apparently in most. Alternatively, they might act as nitrogenous excretory products in the way that urea and uric acid are utilized by animals and birds. Finally, many of the compounds are poisonous, or at least very bitter, and may protect the plant or microorganism from the attentions of herbivores. Little experimental support for these and other suggestions has been obtained, and they must be considered speculative; but the fact remains that many of these compounds probably have vital, as yet unknown roles; these ideas are explored more fully in Chapter 7.

We shall first of all consider those pathways that give rise to the alkaloids, and then discuss other, less common metabolic transformations.

Pathways producing alkaloids

The interesting pharmacological properties of alkaloids have always fascinated Man: the utility of ephedrine, strychnine, cocaine, and other alkaloids was mentioned in Chapter 1. One further example provides an illustration of the crudity of some of these preparations. An ancient Egyptian medical manual contains a remedy for colic in babies: a concoction prepared from poppy pods and fly 'dirt'. The efficacy of such a potion is hardly surprising in light of the pharmacological effects of morphine: analgetic and intestinal calmative.

Such folklore remedies no doubt excited the interest of the nineteenth-century chemists, and provided the incentive for their investigations. Between 1817 and 1832, the alkaloids morphine, strychnine, quinine, nicotine, codeine, and many others were isolated in the pure state. Structure elucidation was often painfully slow; but structural patterns began to emerge, and biosynthetic pathways were proposed after careful consideration of the likely precursors available to the plants. Recent tracer studies have demonstrated the accuracy of many of these intuitive proposals, and we can now state, with certainty, that most alkaloids are derived via the metabolism of the aliphatic amino acids ornithine [252] and lysine [253], and the aromatic amino acids phenylalanine [210], tyrosine [211], and tryptophan [254], all of L-configuration ((S)-stereochemistry). We can further sub-divide these groups by structural type into small families of alkaloids, and these will be discussed in turn.

Alkaloids derived from metabolism of ornithine and lysine

The major structural types are shown in Fig. 5.2. They each contain either C_4N or C_5N sub-units and these are formed from ornithine and lysine respectively.

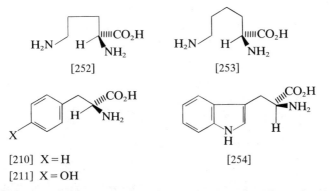

[252]   [253]

[210] X = H
[211] X = OH

[254]

The C_5N sub-unit is often derived from lysine in a specific fashion. That is, when mono-[^{14}C]-labelled lysines are fed, only one carbon atom of the alkaloid is labelled (e.g. eqn 5.1). This was taken as evidence for the non-involvement of the symmetrical diamine, cadaverine [255].

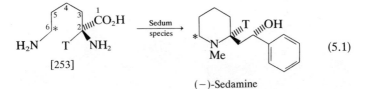

[253]

(5.1)

(−)-Sedamine

(However, cadaverine is present in many plants, and labelled cadaverine has also been incorporated non-randomly into several piperidine alkaloids.) The non-equivalence of the two nitrogen atoms of lysine (at least in some systems) was demonstrated by feeding experiments with

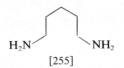

[255]

[6-^{15}N]-lysine and [2-^{15}N]-lysine: the nitrogen atom attached to C-6 was retained, but the atom joined to C-2 was lost. Finally, the involvement of the potential intermediates 2-oxo-6-aminohexanoic acid [256], and piperideinecarboxylic acid [257], is unlikely in light of the tracer results shown in eqn (5.1) (no loss of tritium).

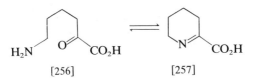

[256]   [257]

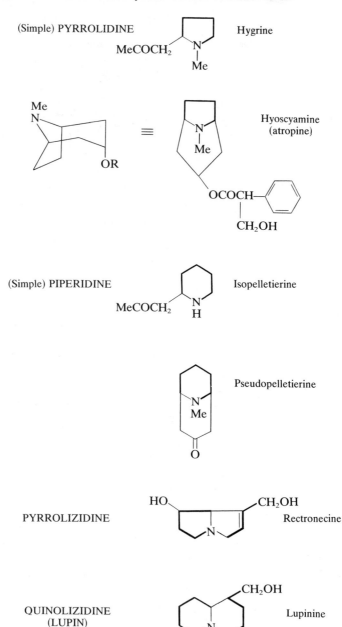

(Simple) PYRROLIDINE — Hygrine

Hyoscyamine (atropine)

(Simple) PIPERIDINE — Isopelletierine

Pseudopelletierine

PYRROLIZIDINE — Rectronecine

QUINOLIZIDINE (LUPIN) — Lupinine

Fig. 5.2

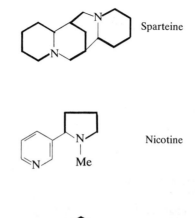

Sparteine

PYRIDINE

Nicotine

Anabasine

Fig. 5.2 (*continued*)

The pathway shown in Fig. 5.3 (due to Spenser) is in accord with all of the evidence, including the specific incorporation of cadaverine. Decarboxylation, and transamination (i.e. the functional group change >CH—NH$_2$ to >C=N— to —CH=O) proceed via the intermediacy of a lysine-pyridoxal phosphate complex. [Pyridoxal phosphate is a cofactor related to vitamin B$_6$, and is of pivotal importance in amino acid metabolism. As shown in Fig. 5.3 the cofactor occurs in two forms: pyridoxal-5′-phosphate (aldehyde form), and pyridoxamine (amine form). In general, it mediates a reversible interconversion of α-amino acids and α-keto acids via a Schiff-base intermediate. See also Staunton (1978).] In those instances where a symmetrical labelling pattern is obtained, it is only necessary to postulate that the equilibrium for cofactor-bound cadaverine to free cadaverine is biased in favour of free cadaverine, thus favouring randomization of label in this symmetrical molecule.

It is likely that the C$_4$N sub-unit is derived from ornithine [252] via the diamine putrescine [258] in a similar fashion (eqn 5.2).

Often N-methyl groups are present, and these are derived from S-adenosyl methionine or from methanoate. The point at which the methyl group is introduced varies according to the pathway concerned.

The actual alkaloid precursors are almost certainly 4-amino-butanal [259], and 5-amino-pentanal [260], which are probably in equilibrium with the cyclic imines, Δ^1-pyrrolideine [261], and Δ^1-piperideine [262]. Simple chemical reactions (enzyme-mediated) such as Schiff-base formation, Mannich condensation, and aldol-type processes, are surely

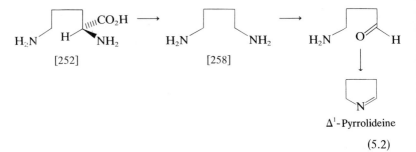

Δ^1-Pyrrolideine

(5.2)

involved in further elaboration of these species, and typical mechanisms are shown in Fig. 5.4.

Almost all alkaloidal structures can be rationalized through application of such processes. Indeed biomimetic total syntheses of many alkaloids

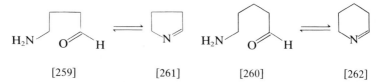

were carried out using these kind of reactions, before the biosynthetic routes were established. Further modifications due to oxidation, hydrogenation, isomerizaton, etc., occur frequently, and are often carried out at a site remote from that at which the primary skeleton is assembled. Thus hyoscyamine [263] is produced in the roots of one species of plant (*Datura*) and not in the leaves. Its 6,7-epoxide, scopolamine (used as a premedication prior to surgery, and as a powerful hypnotic in the treatment of 'the DTs') [264], is only produced in young leaves. In mature

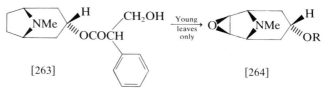

plants [2-14C]-ornithine is converted into hyoscyamine, but not into scopolamine, In other cases, and nicotine [265] is a prime example, the alkaloid is synthesized in the roots of the tobacco plant (*Nicotiana tabacum*), but is then translocated to the leaves and stored.

[265]

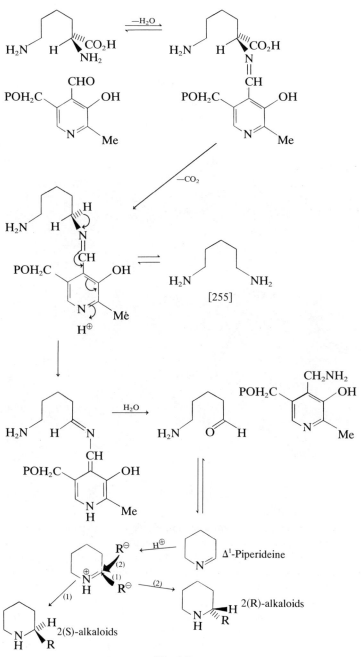

Fig. 5.3

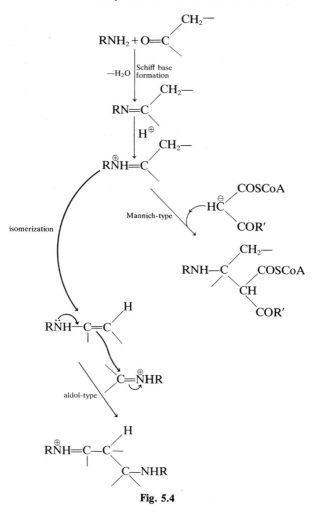

Fig. 5.4

Pyrrolidine alkaloids

The pyrrolidine alkaloids include simple alkylated pyrrolidines such as hygrine [266], as well as the more complex tropane alkaloids like hyoscyamine [263], cocaine [267], and tropine [268]. They are probably formed via alkylation of an N-methylpyrrolinium cation (or biological equivalent), followed by side-chain modification, and subsequent cyclization in the case of tropane alkaloids (Fig. 5.5).

The pathway giving rise to the tropane skeleton is well established:

(1) $[2^{14}C]$-ornithine is incorporated into tropine and hyoscyamine, and only one carbon is labelled, i.e. free putrescine [258] is probably not

involved, cofactor-bound putrescine being the favoured species (see Fig. 5.3).

(2) As expected, experiments utilizing labelled acetate established that only carbon atoms 2, 3, and 4 of [263] are derived from acetate.

(3) Incorporation of [1-^{13}C, methylamino-^{15}C]-N-methylputrescine into scopolamine [264] provided further confirmation of the non-involvement of free putrescine, since only one ^{13}C–^{15}N coupling was observed in the ^{13}C-n.m.r. spectrum, i.e. only one set of contiguous isotopes was present. This also rules out degradation of N-methylputrescine prior to incorporation.

The mode of biosynthesis of the side-chain of hyoscyamine, so-called tropic acid [269], is still unknown. It is derived from phenylalanine [210], but tracer studies have revealed the existence of an intriguing rearrangement (eqn 5.3) which results in formation of tropic acid, and the mechanism of this process has yet to be established.

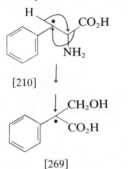

[210]

[269]

(5.3)

Extracts of *Datura stramonium*, *Atropa belladonna*, and *Hyoscyamus niger*, all of which contain tropane alkaloids, have been used for centuries to evoke prophecies, as poisons, and as important adjuncts to religious ceremonies including witchcraft rites. The priestesses of the Delphic Oracle made their prophecies while intoxicated with fumes emanating from burning seeds of henbane (*Hyoscyamus niger*), while witches smeared themselves under the arms ('and in other hairy places') with mixtures of nightshade (*Atropa belladonna*) and fat, the latter to aid absorption through the sebaceous tissues. In a more sinister vein, extracts of mandrake (*Mandragora officianalis*) were used as poisons by the Borgias and their kind, and this plant also contains tropane alkaloids.

Of the tropane alkaloids, cocaine [267] is probably the most interesting from a pharmacological point of view. It derives from the Coca plant (*Erythroxylon coca*), a native of the Andes; and the leaves of this plant are dried, and chewed by the local Indians. An estimated 8 000 000 Indians consume quantities of cocaine daily and they are said to derive a feeling of well-being, and alleviation of hunger pangs, with no concomitant

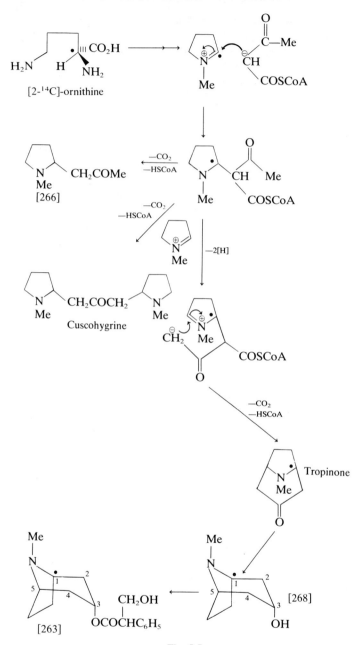

Fig. 5.5

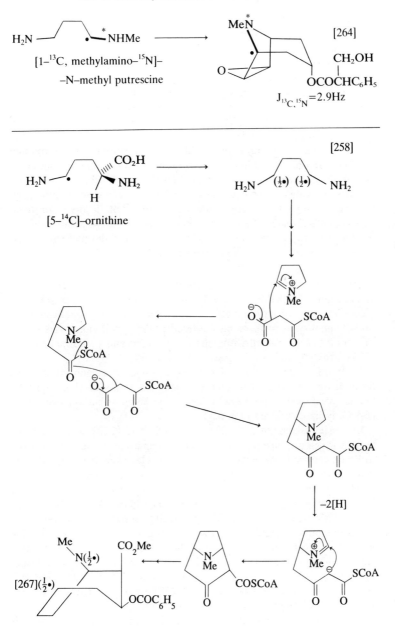

Fig. 5.5 (*continued*)

hallucinogenic effects. However, intravenous cocaine is stongly addictive, and this compound is treated as a narcotic, though it is a useful local anaesthetic, used particularly in minor ear, nose, and throat operations. The biosynthesis of cocaine in *Erythryloxon coca* is apparently different from that of the other tropane alkaloids, and a symmetrical labelling pattern is obtained. Putrescine [258] is the likely symmetrical inter-mediate, and extensive studies by Leete have shown that the probable pathway is different from that proposed for the biogenesis of the other tropane alkaloids (Fig. 5.5).

It is interesting to note that, with rare exceptions, the pyrrolidine alkaloids are only found in one family of higher plants: the Solanaceae, a fact of obvious taxonomic importance. Further, both simple pyrrolidines and tropane alkaloids co-occur in several species, e.g. cuscohygrine (Fig. 5.5) and tropine [268] are major constitutents of deadly nightshade (*Atropa belladonna*). These facts support the biogenetic relationship of the two structural types, as well as re-emphasizing the quite restricted occurrence of certain metabolic pathways.

Piperidine alkaloids

Many higher plants possess the necessary metabolic pathways for the elaboration of alkaloids from lysine. In each case these alkaloids incorporate one or more piperidine rings. The biosynthetic reactions are broadly analogous to those encountered in the pyrrolidine series. Thus, the formation of N-methylisopelletierine [270], and pseudopelletierine [271], probably follows the course shown in eqn (5.4).

Many of these alkaloids incorporate lysine in a non-symmetrical fashion: we have already noted the tracer work with sedamine (eqn 5.1). Similarly, (−)-halosaline [272] is labelled at only one carbon atom when [6-^{14}C]-lysine is utilized (eqn 5.5).

In passing we might note that L-pipecolic acid [273] occurs in many species which produce piperidine alkaloids, and was at one time believed to lie on the same pathway of metabolism. However, it is formed from D-lysine more efficiently than from L-lysine (the precursor of all of the alkaloids), and is probably derived via Δ^1-piperideine carboxylic acid [257], which is not on the pathway to the piperidine alkaloids (eqn 5.6).

Labelling experiments have also provided an insight into the biogenesis of several more complex alkaloids which derive from lysine. A good example is lycopodine [274], one of the so-called Lycopodium alkaloids (from Club mosses). Two molecules of lysine are incorporated, and since symmetrical incorporation of label is observed, free cadaverine is a probable intermediate. Isopelletierine [270] is also incorporated, and the remaining carbon atoms are of polyketide origin.

At this stage it is worth mentioning the hemlock alkaloids, notably (+)-coniine [275], and γ-coniceine [276], since although they bear an obvious structural resemblance to the typical piperidine alkaloids, they

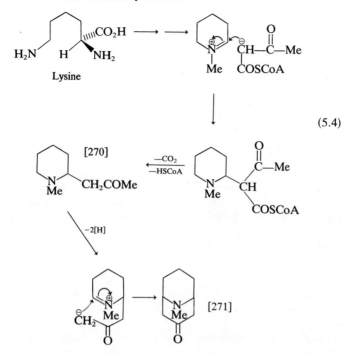

(5.4)

are derived via a totally different pathway: another example of the diversity of biosynthetic pathways leading to similar structural types. The toxicological effect of hemlock has been known for a very long time, and has been responsible for the demise of many celebrities (Socrates for

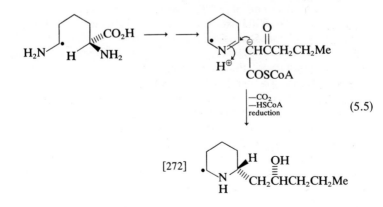

(5.5)

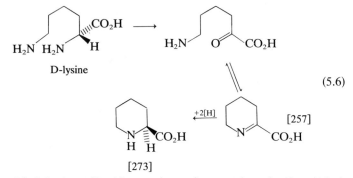

(5.6)

[273]

example). It is also utilized by certain carnivorous plants for the paralysis of trapped insects. Coniine was also the first alkaloid to be synthesized (1886), and due to structural similarities to isopelletierine [270], a

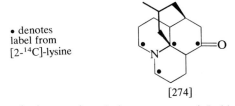

• denotes
label from
[2-^{14}C]-lysine

[274]

probable biosynthetic route from lysine was assumed. In hindsight (and it is always easy to be wise after the event) taxonomic considerations might have led to suspicions that all was not well, since the plant family to which hemlock is assigned (Umbelliferae) is only distantly related to those plant families which contain typical piperidine alkaloids. In consequence, a similar biosynthetic pathway would be unlikely.

In fact, isotopically labelled lysine was poorly incorporated, and subsequent experiments have established a polyketide origin for these metabolites. The probable pathway is shown in Fig. 5.6. Specific incorporations of acetate, of [6-^{14}C]-5-oxo-octanoic acid [277] (* denotes label), and of [6-^{14}C]-5-oxo-octanal [278], were accomplished: the intermediacy of the 5-oxo-compounds was firmly established by the isolation of labelled [277] when [1-^{14}C]-octanoic acid was utilized. The enzyme responsible for the transamination, (L)-alanine: 5-keto-octanal aminotransferase, has been isolated and characterized.

Similarly, nigrifactin [279], a metabolite produced by a strain of *Streptomyces*, is derived entirely from acetate (eqn 5.7).

[1-^{13}C]-acetate →
(•)

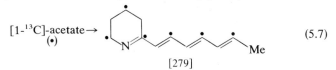

(5.7)

[279]

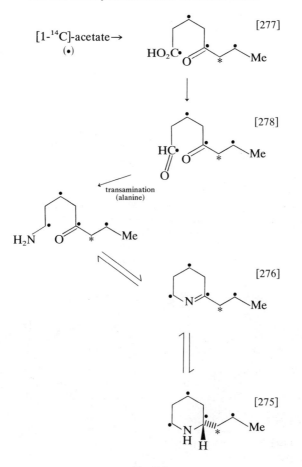

Fig. 5.6

These results provide a salutary reminder of the dangers inherent in assuming that similar structural types are formed by similar metabolic pathways.

Pyrrolizidine alkaloids
This class of alkaloids, together with the quinolizidine alkaloids (next section), are best considered separately although they are derived from ornithine and lysine, respectively, via similar biosynthetic pathways. In both instances a symmetrical intermediate has been implicated since the two carbon atoms, adjacent to the nitrogen atom, become equivalent. The pyrrolizidine alkaloids usually occur as esters of unique mono- or dibasic

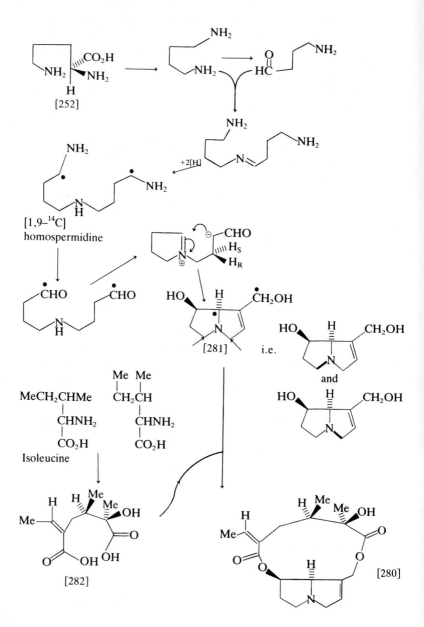

[252]

$[1,9{-}^{14}C]$
homospermidine

[281] i.e.

and

Isoleucine

[282]

[280]

Fig. 5.7

acids: the so-called necic acids. Thus, senecionine [280] is derived from retronecine [281] and senecic acid [282]. A possible route to retronecine has been proposed by Robins (Fig. 5.7), on the basis of labelling studies with putrescine and homospermidine. Statellites were visible in the ^{13}C-spectrum of [281] when [1-amino-^{15}N, 1-^{13}C]-putrescine was used, and these were assigned to the carbon atoms arrowed. This clearly demonstrates the involvement of a symmetrical intermediate. The later stages have been investigated using 2(R) and 2(S) mono-deuterated putrescines, and it was shown that the hydroxylation occurred with retention of configuration, and that the double bond was introduced with loss of the pro-2(S) hydrogen. In addition, Crout has established that senecic acid is produced by condensation of two molecules of isoleucine.

It is worth noting that many pyrrolizidine alkaloids exhibit dramatic hepatic toxicity, and livestock are often poisoned. It is believed that the toxicity is due to pyrrole esters of the type [283], which have potent alkylating capabilities. These esters are produced by metabolism of pyrrolizidine alkaloids in the mammalian liver.

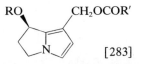

[283]

In addition these alkaloids are believed to have an ecological role in at least two species of butterflies. The larvae of the cinnabar moth accumulate such alkaloids in their tissues by feeding on groundsel or ragwort (both rich in pyrrolizidine alkaloids), and thus become distasteful to predators. Certain butterflies of the sub-family Danaid utilize pyrrolizidine alkaloids as precursors for the production of dehydropyrrolizidines, such as [284]. These are used by the male butterflies as female flight arrestants or as aphrodisiacs, and thus play a vital part in the prelude to mating.

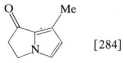

[284]

Quinolizidine alkaloids

These alkaloids are sometimes known as the Lupin alkaloids, e.g. lupinine [285], since they occur widely, but not exclusively, in species of the genus *Lupinus*. Their biogenesis from lysine is perhaps analogous to that of the pyrrolizidine alkaloids and careful labelling studies by both Robins and Spenser yielded the pattern shown in Fig. 5.8. Equal levels of ^{13}C-enrichment were observed for six carbon atoms, and there were two ^{13}C–^{15}N doublets (3.7 and 3.4 Hz). This demonstrates that two moles of

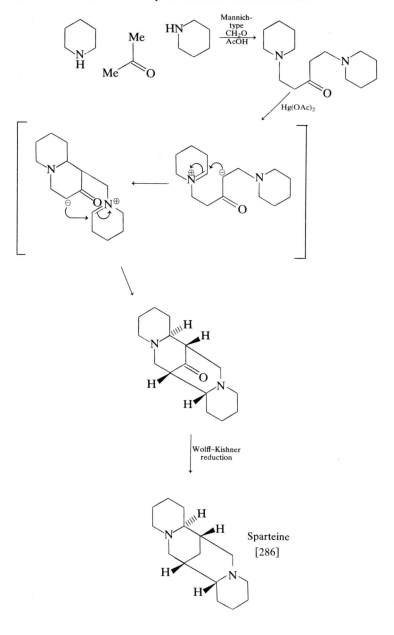

Fig. 5.8

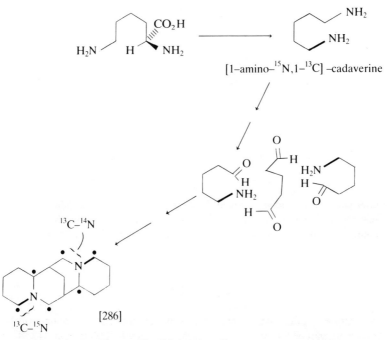

$[1-amino-{}^{15}N,1-{}^{13}C]$ –cadaverine

Fig. 5.8 (*continued*)

cadaverine are incorporated specifically into the two outer rings of sparteine [286], and Robins has suggested that two moles of 5-aminopentanal and one mole of pentanedial are involved, probably in the form of enzyme-linked intermediates. Spenser has interpreted the results in terms of the intermediacy of a modified trimer of piperideine [262], and has proposed a reasonable though complex biogenetic pathway. Late-stage intermediates will have to be isolated before we can verify these suggestions.

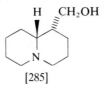

[285]

Possibly the commonest quinolizidine alkaloid is sparteine [286], and a particularly good biomimetic total synthesis of this compound has been carried out by Van Tamelen. This is also shown in Fig. 5.8. It is interesting to note that the synthesis of sparteine by the broom plant seems to be of negative survival value, since it stimulates attack by aphids. Presumably aphids have evolved in such a way as to overcome what at one time was a chemical means of deterrence; and this is a common theme underlying secondary metabolism and plant–insect interactions.

Pyridine alkaloids
The biogenesis of pyridine alkaloids, such as nicotine [265] and anabasine [287] (both tobacco alkaloids) is conceptually quite simple. The C₄N and

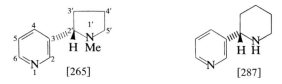

$$C_4N$$

C₅N sub-units could be derived from ornithine [252] and lysine [253], respectively; and nicotinic acid [288] or an enzyme-bound thiol ester could serve as the precursor of the pyridine ring. [2-^{14}C]-ornithine is

incorporated into nicotine by tobacco plants, and yields a symmetrical labelling pattern: positions 2′ and 5′ are labelled. Putrescine [258], N-methylputrescine, and N-methylaminobutanal [289] are all incorporated, and quite recently the individual enzymes that catalyse some of the steps in the biosynthetic pathway have been isolated (Fig. 5.9). The intermediacy of the N-methylpyrrolinium ion is supported by recent experiments in which a variety of such compounds were converted into analogues of nicotine, by *Nicotiana glutinosa* (eqn 5.8).

The overall biosynthetic process is thus an electrophilic aromatic substitution at C-3 of a pyridine ring: a predictable finding, in light of the known reactivity of the pyridine nucleus.

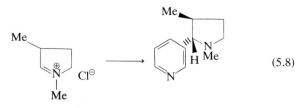

$$(5.8)$$

In contrast to the symmetrical labelling pattern observed in nicotine biosynthesis, radio-labelled lysine is incorporated into anabasine [287] in a non-symmetrical fashion. However, the structurally similar anatabine [290] is not formed from lysine, and appears to be derived entirely from

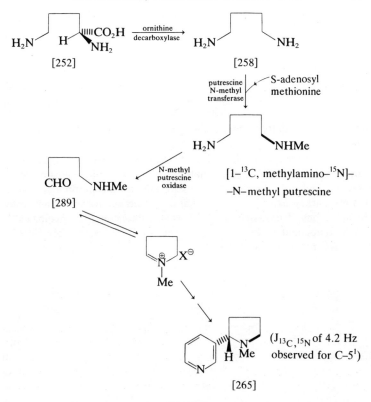

Fig. 5.9

nicotinic acid [288], illustrating once more the capriciousness of secondary metabolism.

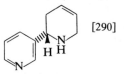

[290]

As to the biogenesis of the pyridine ring: since quinolinic acid [291] is as efficiently incorporated as nicotinic acid, a pathway via successive decarboxylations is implicated. Quinolinic acid is formed from glyceraldehyde-3-phosphate [292] (a product of glucose metabolism), and aspartic acid [293]; and a plausible mechanism is shown in eqn (5.9). It should be noted that although this metabolic pathway is utilized by higher plants, and certain anerobic bacteria and yeasts, it is not used by higher organisms (animals and aerobic microorganisms). They synthesize

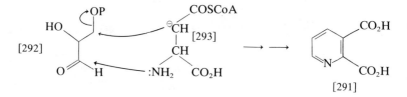

nicotinic acid from tryptophan [254] via-3-hydroxy-anthranilic acid [294] (Fig. 5.10).

The final stages of nicotine biosynthesis can be envisaged to occur as shown in eqn (5.10). We might note in passing that nicotine is a useful insecticide, and several hundred tonnes are used for this purpose annually: it is not unreasonable to suppose that it protects the tobacco plant from the attentions of browsing insects. It is also the main pharmacologically active component of tobacco smoke, and probably responsible for the 'addictive' nature of cigarettes.

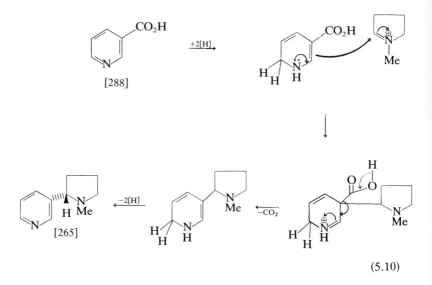

(5.10)

Finally, one other metabolite should be mentioned in this section: ricinine [295]. This is a toxic alkaloid from the 'castor oil' plant (*Ricinus communis*), and is derived from nicotinic acid via nicotinamide [296]. A crude enzyme preparation has been isolated, and this will convert a series of pyridinium salts into ricinine and related compounds. A plausible biosynthetic scheme is shown in Fig. 5.11; the observed reversible demethylation/methylation is particularly noteworthy.

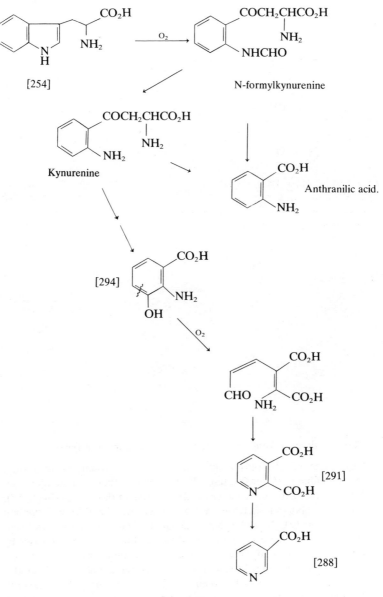

Kynurenine

N-formylkynurenine

Anthranilic acid.

[254]

[294]

[291]

[288]

Fig. 5.10

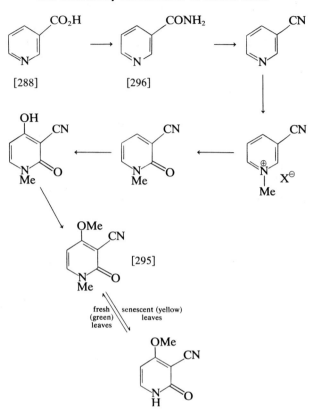

Fig. 5.11

Alkaloids derived from metabolism of phenylalanine and tyrosine

This group of alkaloids contain an $ArylC_2N$ sub-unit derived from phenylalanine [210] or tyrosine [211], often together with an additional ArC_2 or ArC_1 sub-unit from partial degradation of these acids (Fig. 5.12). The basic alkaloid skeletons are produced via Schiff-base formation between an ArC_2N unit (phenylethylamine) and an 'active' carbonyl component (aliphatic or aromatic). Modification of the initial species by decarboxylation, transamination, imine formation, hydroxylation, or, most importantly, by oxidative phenolic coupling (usually C—C but also C—O coupling) gives rise to a seemingly bewildering array of compounds, some of them very complex. However, once again the chemistry involved is usually fundamental Schiff-base formation and phenolic coupling are the major structure-determining reactions.

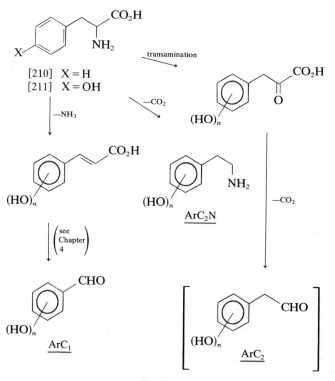

[210]　X = H
[211]　X = OH

transamination

−CO₂

−NH₃

(HO)ₙ

ArC_2N

−CO₂

(HO)ₙ

$\left(\begin{array}{c}\text{see}\\\text{Chapter}\\4\end{array}\right)$

ArC_1

ArC_2

Fig. 5.12

Four main structural types may be distinguished:
(1) simple monocyclic compounds;
(2) Isoquinolines;
(3) benzylisoquinolines;
(4) Amaryllidaceae alkaloids.

Some representative examples are shown in Fig. 5.13: the biogenetic sub-units have been emphasized. It should be noted that the terms isoquinoline and benzylisoquinoline are merely simple generic names, and do not imply the level of oxidation of the system: the systems are usually tetrahydro- or dihydro-(benzyl)isoquinolines.

Simple monocyclic compounds
It is well established that in barley (*Hordeum vulgare*) phenylalanine is converted into tyrosine and thence, via tyramine [297] and N-methyl tyramine into hordenine [298] (eqn 5.11). As noted previously, alkaloids are usually derived from phenylalanine *or* tyrosine, since the ability to convert the former into the latter is confined mainly to the grasses: barley

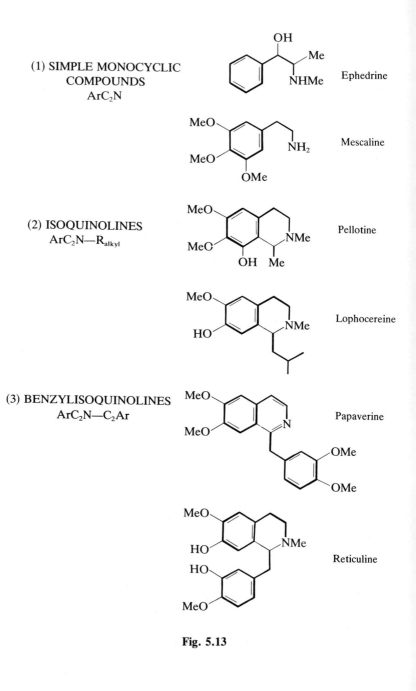

(1) SIMPLE MONOCYCLIC
COMPOUNDS
ArC_2N

Ephedrine

Mescaline

(2) ISOQUINOLINES
ArC_2N—R_{alkyl}

Pellotine

Lophocereine

(3) BENZYLISOQUINOLINES
ArC_2N—C_2Ar

Papaverine

Reticuline

Fig. 5.13

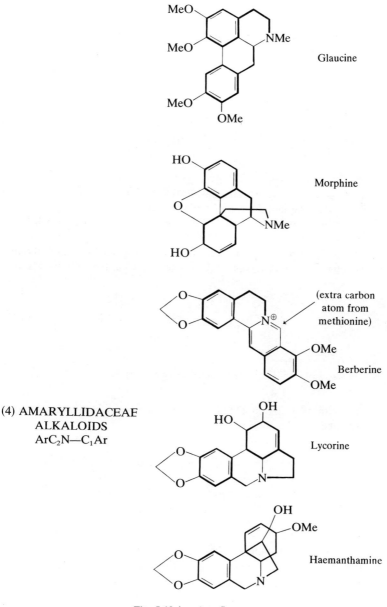

Glaucine

Morphine

(extra carbon atom from methionine)

Berberine

(4) AMARYLLIDACEAE ALKALOIDS
ArC$_2$N—C$_1$Ar

Lycorine

Haemanthamine

Fig. 5.13 (*continued*)

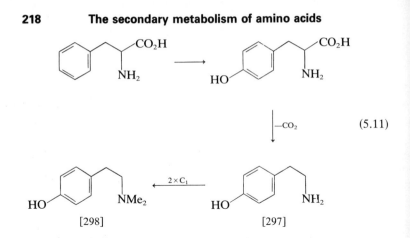

(5.11)

[298]   [297]

has this capability. Interestingly, two quite discrete N-methylases are involved in the methylation of tyramine in barley.

A related species, *Hordeum distichum*, converts labelled phenylalanine into N-methyl tyramine with the interesting labelling pattern shown in eqn (5.12). This is a good example of the so-called NIH shift (the first example of such an isotopic shift was discovered at the National Institute

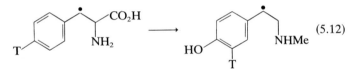

(5.12)

(88% retention of tritium)

of Health, Washington, D.C.). The reaction is believed to proceed via the intermediacy of a species like [299].

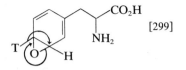

[299]

Another well-studied alkaloid is ephedrine [300], a useful bronchodilating agent, employed in asthma therapy; it also stimulates the central nervous system and elevates blood pressure, and must be used with care. The biosynthetic pathway is of interest and appears to proceed via the route shown in eqn (5.13).

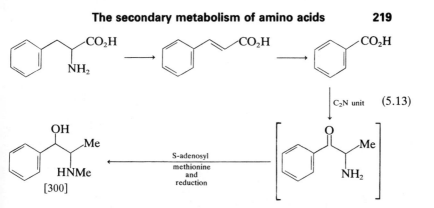

(5.13)

[300]

A series of simple monocyclic (as well as isoquinoline-type) compounds are produced by a variety of cactus species. The major psychoactive component of the 'peyote' cactus (*Lophophora williamsii*) is mescaline [301]. This has interesting hallucinogenic properties, and is used as an integral part of religious activities by thousands of American and Mexican indians. Mescaline is derived from dopamine (3,4-dihydroxyphenylethylamine) [302] as shown in eqn (5.14).

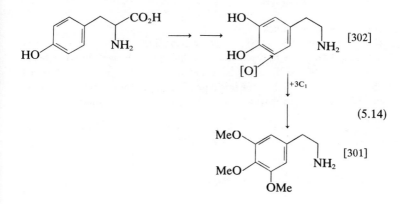

(5.14)

Two somewhat obscure alkaloids should also be considered at this point: gliotoxin [303], a fungal antibiotic and antiviral agent (but too toxic for clinical use) and betanin [304], the red pigment of beetroot. Gliotoxin is derived from phenylalanine and serine [305], and an epoxide intermediate has been suggested (eqn 5.15) to account for the lack of involvement of *m*-tryosine. Betanin is almost certainly derived from 3,4-dihydroxyphenylalanine (DOPA) [306], and a reasonable biosynthetic pathway is shown in Fig. 5.14.

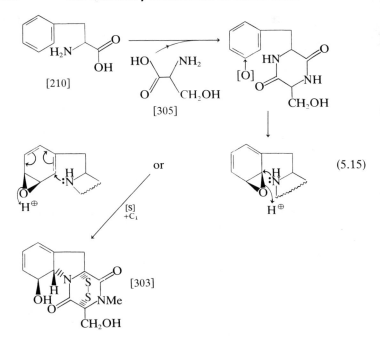

(5.15)

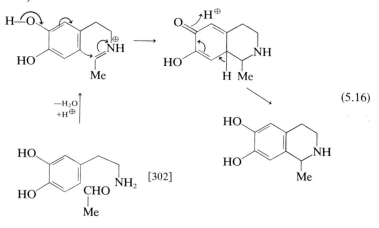

Isoquinoline compounds

The isoquinoline (or rather, tetrahydroisoquinoline) ring system can be considered to arise by cyclization of the Schiff base formed between dopamine [302] and an aliphatic aldehyde (or biological equivalent). There is ample precedent for this type of reaction, for example, dopamine and ethanal react, *in vitro* at pH 5, to yield a tetrahydroisoquinoline (eqn 5.16).

(5.16)

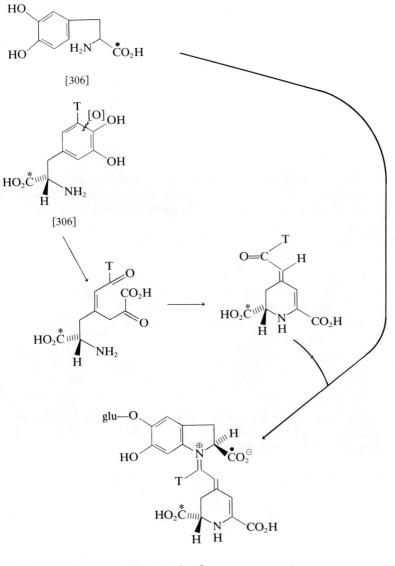

[306]

[306]

[304]

Fig. 5.14

In the case of the simplest isoquinoline alkaloid, anhalamine [307], methionine supplies the formaldehyde equivalent (eqn 5.17). The other common aliphatic unit is ethanal, derived in all probability from pyruvate

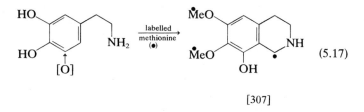

(5.17)

[307]

as shown for the biogenesis of pellotine [308] (eqn 5.18); pellotine and anhalamine are both obtained (like mescaline) from the 'peyote' cactus. A peyote-O-methyl transferase has been isolated and characterized.

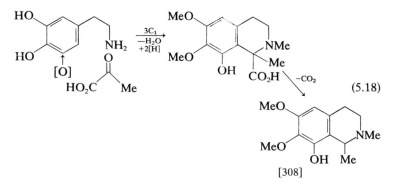

(5.18)

[308]

Other aliphatic sub-units are encountered occasionally, and the C_5 unit of lophocereine [309] is probably derived from the keto acid [310], itself almost certainly formed from MVA, or from the amino acid leucine.

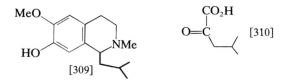

[309]   [310]

As usual there is some doubt as to the timing of the various reactions: hydroxylation, methylation, cyclization, etc. In fact the actual sequence of events probably varies somewhat from species to species, even when structurally similar metabolites are produced. For example, recent work on cell-free systems from the 'peyote' cactus has shown that the alkaloids [307] and [308], which co-occur, are formed via quite different metabolic pathways; and it is probable that this finding is of wide generality in the alkaloid field.

Benzylisoquinoline compounds
These metabolites can be divided into three basic structural types: benzylisoquinolines (strictly tetrahydro-benzylisoquinolines) with no structural alteration due to oxidative phenolic coupling, and those produced as a result of two primary modes of coupling (Fig. 5.15). The benzylisoquinoline ring system is produced via cyclization of Schiff bases formed from dopamine [302] (or an equivalent species) and aryl aldehydes. Thus, norlaudanosoline [311] and papaverine [312] are derived as shown in Fig. 5.16. The involvement of 3,4-dihydroxphenylpyruvates was originally suggested to explain the fact that dopamine is only incorporated into the isoquinoline portion of these compounds. The aryl pyruvate can be derived from tyrosine, but not from DOPA [306] or from dopamine. The utilization of the requisite aldehyde, at least for the production of [311], has been established by incorporation studies.

The alkaloid reticuline [313] is formed in a similar fashion, and occurs widely in Nature. The (S)-(+)-enantiomer is most commonly encountered, but the (R)-(−)-enantiomer is the main precursor of the opium alkaloids including morphine.

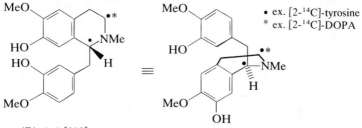

• ex. [2-^{14}C]-tyrosine
* ex. [2-^{14}C]-DOPA

(R)–(−)-[313]

A minor modification of the isoquinoline skeleton produces alkaloids like berberine [314]. An extra carbon atom, from methionine, is incorporated into the skeleton. In all probability the N-methyl group is transformed into an imine, which suffers attack in the usual way, and oxidation completes the synthesis (eqn 5.19); and the enzymes responsible for each step have been isolated and characterized.

Before discussing the two main structural types produced by oxidative phenolic coupling, it is worth emphasizing how much the inductive approach has contributed to our understanding of these biosynthetic pathways. Long before tracer experiments were carried out, the presence of the dihydroxyphenylethylamine sub-unit [302] within these alkaloids was recognized, and possible biosynthetic pathways were induced.

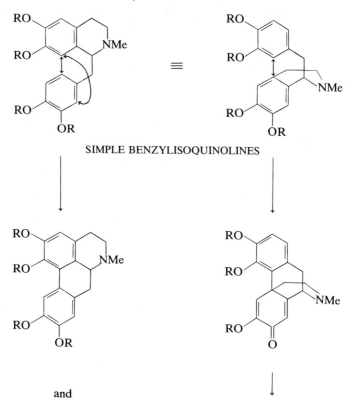

SIMPLE BENZYLISOQUINOLINES

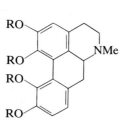

and

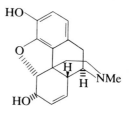

OPIUM ALKALOIDS

(R = H or Me)
APORPHINE ALKALOIDS

e.g. Morphine

Fig. 5.15

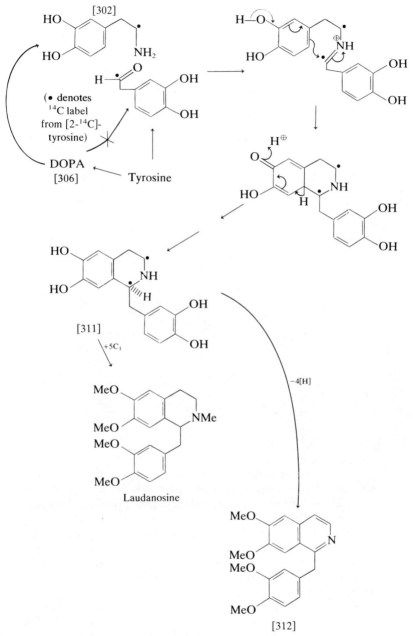

Fig. 5.16

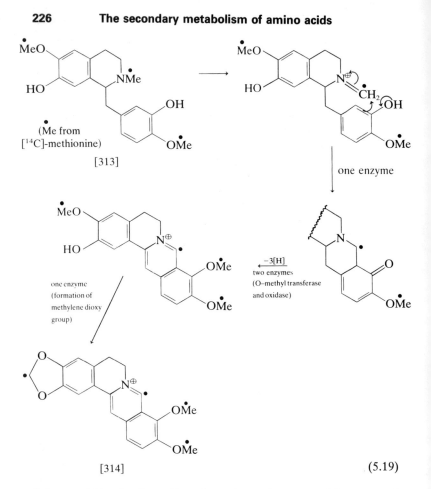

[314]   (5.19)

Subsequent biomimetic syntheses have shown that many of the proposed intermediates can be prepared *in vitro*, and that most of the postulated interconversions are feasible. It is only quite recently that corroborative tracer evidence has been obtained, and the original hypotheses have been largely substantiated.

(R)-(−)-Reticuline [313] is the progenitor of the opium alkaloids, and the opium poppy (*Papaver somniferum*) is probably the best source of these compounds. Tracer evidence has firmly established the sequence shown in Fig. 5.17, which includes a C—C coupling step:

(1) [2-¹⁴C]-tyrosine and [1-¹⁴C]-dopamine are incorporated to produce morphine [315] labelled at C-9 and C-16, and at C-16, respectively;

(2) reticuline is incorporated into thebaine [316] as was salutaridine [317] (only found in trace quantities in the opium poppy), and the final

sequence from thebaine to morphine was established by time versus incorporation studies with $^{14}CO_2$.

A discussion of the opium alkaloids would not be complete without some mention of the pharmacological properties of these compounds. The use of opium predates the dawn of historical records, but was mentioned by Homer in the 'Odyssey'. As defined, opium is the air-dried, milky exudate from unripe fruit capsules of the opium poppy. Crude opium contains ca. 10 per cent by weight of morphine [315] and 25 per cent by weight of total alkaloids. Morphine is undoubtedly the major active constituent, and has marked analgetic and narcotic effects. It also produces a constipative effect on the intestinal tract (hence its use as a paregoric), and prolongation of this effect, together with its addictive nature and depressive action on respiration, are the main reasons for its toxicity when abused. Codeine [318] is much less potent and less addictive, while heroin diacetylmorphine, [319] is just the opposite, hence

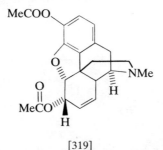

[319]

its wide abuse throughout the Western world. About 85 per cent of the heroin reaching Western countries comes from the poppy fields of S. E. Asia and Turkey, via refining centres in Europe. The compound methadone [320] appears to have the necessary structural characteristics for acceptance by the opiate receptors in the brain, and since it is active orally it removes the necessity for intravenous injection. It is also less addictive than heroin, and is the drug of choice for the clinical treatment of drug addiction. Less abusable drugs such as the combination

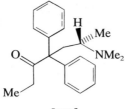

[320]

methadone/naloxone (a synthetic analogue of morphine) are currently under evaluation. Overall, codeine [318], methadone, and morphine [315] account for about 30 per cent of all analgesics used today.

$$H_2N—tyr—gly—gly—phe—met—CO_2H$$

$$H_2N—tyr—gly—gly—phe—leu—CO_2H$$

[321]

Recently 'natural' peptide opiates [321], termed 'enkephalins', have been isolated from the brains of pigs and other mammals including Man. They are derived from larger brain polypeptides (the endorphins), and both types have a marked analgetic effect, and appear to act at the same receptors as the opium alkaloids. Their role is as yet unknown.

In the opium poppy, morphine is rapidly metabolized to normorphine [322] (no methyl groups), which is then degraded. The entire sequence from thebaine [316] to normorphine thus comprises a series of

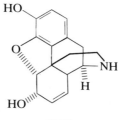

[322]

demethylations, and it has been suggested that these alkaloids may be endogenous alkylating agents. Certainly, the high turnover of the opium alkaloids *in vivo*, eventually producing non-alkaloid metabolites, provides a good example of alkaloids as dynamic metabolites rather than as useless end-products of metabolism.

[2-^{14}C]-tyrosine (▲) $\longrightarrow$ $\longrightarrow$
[1-^{14}C]-dopamine (*)

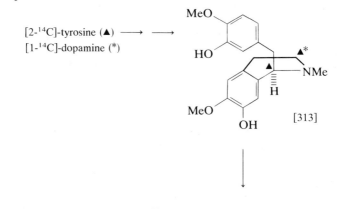

[313]

Fig. 5.17

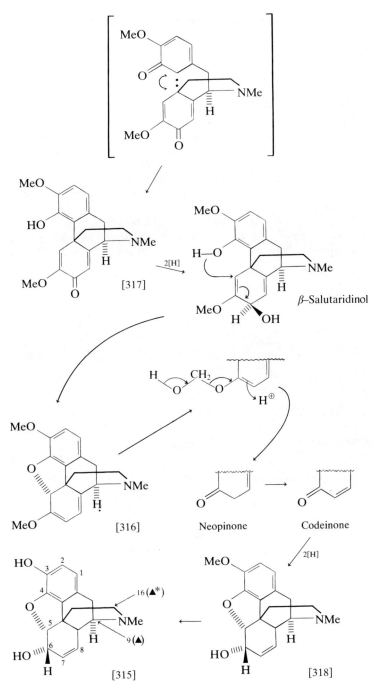

MeO

HO

NMe

H

MeO

O

[317]

$\xrightarrow{2[H]}$

MeO

H—O

NMe

H

MeO

H

OH

β–Salutaridinol

H—O—CH$_2$—O

H$^{\oplus}$

MeO

O

NMe

H

MeO

[316]

Neopinone → Codeinone

$\downarrow$ 2[H]

MeO

O

NMe

H

HO

H

[318]

HO

3

4

2

1

16(▲*)

O

5

NMe

9(▲)

HO

6

8

H

7

[315]

Fig. 5.17 (*continued*)

Finally, several biomimetic syntheses of opium alkaloids have been carried out, and the synthesis of thebaine (eqn 5.20) is exemplary. It is interesting to note that the intermediates are very similar to their natural counterparts.

The alternative mode of oxidative phenolic coupling produces some aporphine alkaloids. Experimental proof has been obtained for the conversion of reticuline [313] into bulbocapnine [323], presumably by direct phenol coupling, and certain similar metabolites are likewise produced from laudanosoline [324] (eqn 5.21). However, the biosynthesis of some other aporphine alkaloids, such as glaucine [325], apparently occurs via a dienone-phenol-type rearrangement, as shown in Fig. 5.18.

A similar pathway, again commencing with norprotosinomenine [326], has been proposed for the biogenesis of the Erythrina alkaloids, e.g. erythraline [327] via the intermediacy of [328] and [329] (see Fig. 5.18).

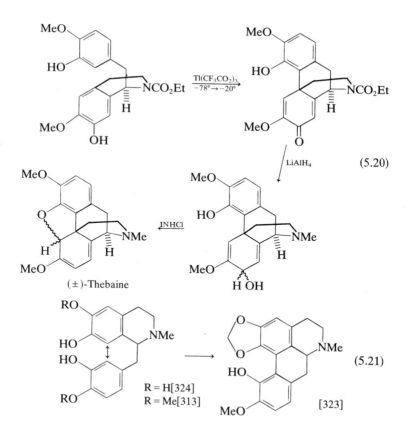

(5.20)

(5.21)

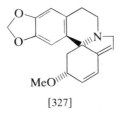

[327]

Amaryllidaceae alaloids

This group of metabolites, which occur mainly in plants of the daffodil family (Amaryllidaceae), are known to derive from O-methylnorbelladine [330]. This in turn is derived from one molecule of phenylalanine [210] and one of tyramine [297], the former being converted into 3,4-dihydroxybenzaldehyde (or the corresponding phenylpyruvate) prior to incorporation (eqn 5.22). The presence of a disubstituted phenol ring means that an extra mode of phenolic coupling can occur, and the possible modes are shown in structure [331]. Once again dienone-phenol-type rearrangements have been implicated, and the three basic processes can

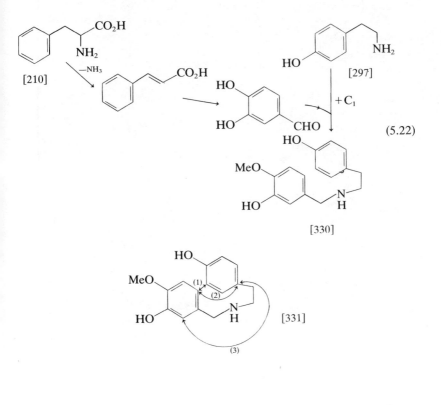

(5.22)

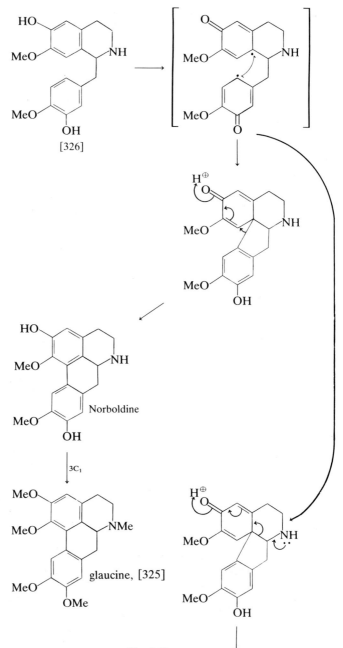

[326]

Norboldine

glaucine, [325]

Fig. 5.18

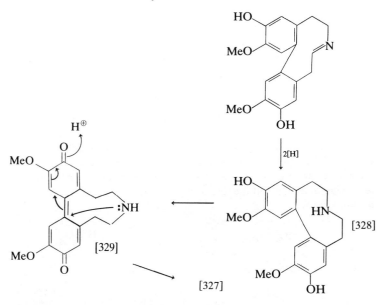

Fig. 5.18 (*continued*)

be considered under the headings: (1) *ortho-para* coupling; (2) *para-para* coupling; and (3) *para-ortho* coupling.

(1) The alkaloids lycorine [332] and norpluviine [333] can be considered to arise via *ortho-para* coupling of O-methylnorbelladine [330] as shown in Fig. 5.19. This sequence is broadly analogous to that involved in the biogenesis of the erythrina alkaloids.

(2) *Para-para* coupling of O-methylnorbelladine may be invoked to explain how the compounds haemanthamine [334] and haemanthidine [335] are formed (Fig. 5.20). Battersby has recently established that there is retention of configuration at the bridgehead carbon atom when the hydroxyl group is introduced: that is, the hydroxyl enters from the side of the molecule from which the hydrogen is displaced. Subsequent hydroxylation of haemanthamine to produce haemanthidine is also stereospecific.

(3) The final mode of coupling involves a rearrangement initiated by electron donation from oxygen, rather than from nitrogen as in the previous processes. Alkaloids like narwedine [336] and galanthamine [337] are produced in this way (Fig. 5.21).

The tracer experiments that have been carried out have produced results which largely substantiate the pathways shown in Figs 5.19–5.21.

Finally, before leaving the subject of alkaloids derived from phenylethylamine units, we should consider briefly the biogenesis of the complex alkaloid colchicine [338]. This is produced by the Autumn

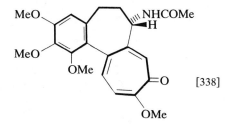

[338]

crocus, and has for centuries been valued as a treatment and diagnostic aid for gout. Relief from pain is produced in over 90 per cent of cases, and its mode of action probably arises from its ability to inhibit the movement of leucocytes and to inhibit the release of enzymes which mediate inflammatory processes.

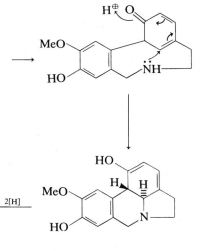

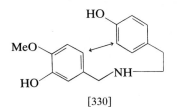

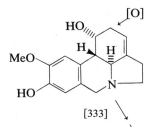

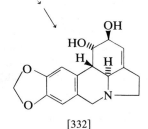

[332]

Fig. 5.19

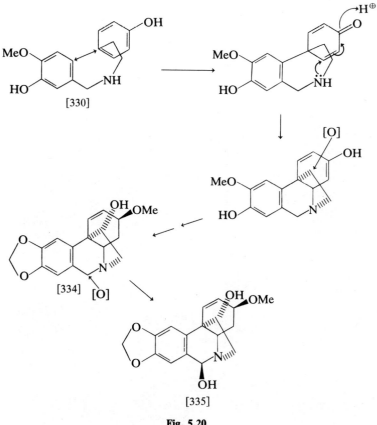

Fig. 5.20

Both phenylalanine and tyrosine are precursors of colchicine: the former supplies the ArylC$_3$ moiety, and the latter provides the tropolone ring and the nitrogen atom. The co-occurrence of androcymbine [339] provided the clue which led Battersby and coworkers to postulate the biosynthetic pathway shown in Fig. 5.22. Good incorporations of the

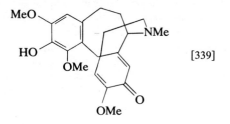

[339]

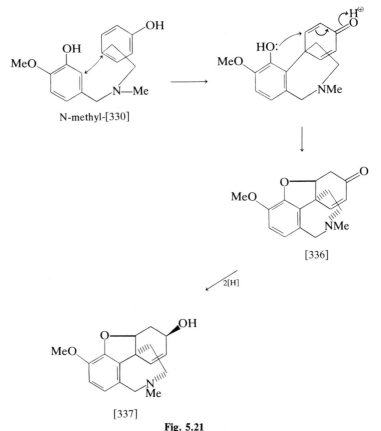

N-methyl-[330]

[336]

2[H]

[337]

Fig. 5.21

advanced intermediates autumnaline (10%) and O-methylandrocymbine (15%) were noted when radio-labelled samples were employed; in addition, [1-^{13}C] autumnaline was synthesized and incorporated. The natural abundance spectrum of colchicine and the enriched spectrum following incorporation of this last compound are shown in Spectrum 5.1 (a) and (b), and it is clear that only one carbon signal has been enhanced. This result accords precisely with the proposed pathway, and the experiment represented something of a milestone in ^{13}C-methodology since it was the first time that an advanced intermediate had been shown to be incorporated solely through the use of ^{13}C-n.m.r. methods. Only the key steps of this rather complex sequence are shown in Fig. 5.22, and for each intermediate the carbon—carbon bonds which derive originally from tyrosine appear in bold type.

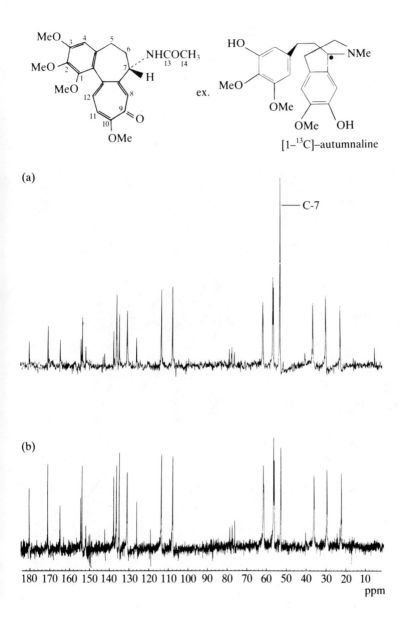

Spectrum 5.1. (from *Tetrahedron Lett.*, 1974, 3315–3318.)

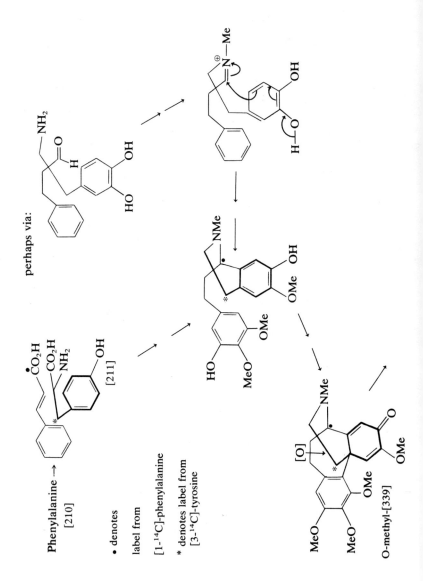

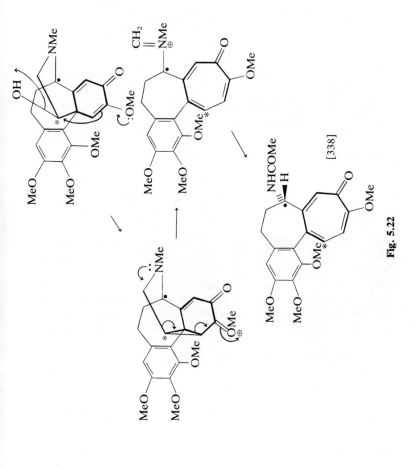

Fig. 5.22

[338]

Alkaloids derived from tryptophan

Metabolism of the amino acid tryptophan [254] provides the indole-C_2N sub-unit found in a wide variety of alkaloids. These compounds fall into two major classes:

(1) simple alkaloids formed by minor chemical modification of tryptophan, or by incorporation of a C_2 sub-unit (from pyruvate) into the structure (this latter type is analogous to the isoquinoline class);

(2) alkaloids produced as a result of mixed metabolism of tryptophan and mevalonate, and thus incorporating a C_5 or C_9/C_{10} sub-unit (the C_9 sub-unit is obtained by modification of a C_{10} unit).

This second class will be considered in the next chapter, but representative examples of both classes are shown in Fig. 5.23.

The reactions of the indole nucleus can be rationalized in terms of the potentially nucleophilic nature of C-3, as depicted in structure [340], and in eqn (5.23). The intermediacy of spiro intermediates, such as [341], has

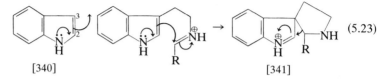

$$(5.23)$$

[340]   [341]

been established in simple *in vitro* systems, and may also occur *in vivo*. The biosynthesis of many of the more complex indole alkaloids has yet to be studied, but reasonable pathways have been proposed based on the known reactivity of the indole nucleus.

The incorporation of radio-labelled tryptophan into simple indole alkaloids has been amply demonstrated. These alkaloids differ primarily in the position of hydroxylation, and in the chemical modifications of the side-chain.

Serotonin [342] is widely distributed in Nature, and is a natural neuro transmitter in the central nervous system, responsible at least in part for control of sleep patterns. Bufotenine [343] is the N,N-dimethyl analogue.

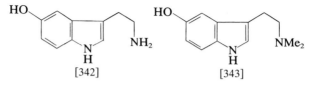

[342]   [343]

There is also some evidence that they are involved in causation of depression and perhaps of schizophrenia. Both of these metabolites have 5-hydroxyl groups. Species of the mushroom *Psilocybe* produce metabolites like psilocin [344] and psilocybin [345] which have 4-hydroxyl groups. These compounds have hallucinogenic properties

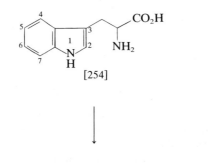

[254]

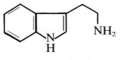

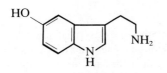

Indole-C_2N

C_5 units

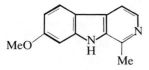

Serotonin

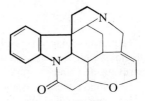

Strychnine

Harmine

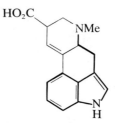

Lysergic acid

Fig. 5.23

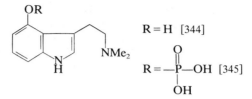

R = H [344]

R = —P—OH [345]

similar to those of LSD, and the 'magical' properties of these mushrooms have been known in Central and South America for at least 3000 years.

A non-oxygenated alkaloid, gramine [346], found principally in germinating barley, is also derived from tryptophan, but the mechanism of side-chain modification is unknown. Recent tracer results indicate that an indole-C_1 sub-unit is incorporated intact, but the origin of the second nitrogen atom is in doubt (eqn 5.24). Gramine appears to act as an allelopathic agent since it inhibits the growth and germination of other plants. This might account for the success of barley in competition with other species, especially other grasses.

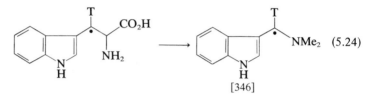

(5.24)

[346]

The alkaloid harmine [347] is representative of a class of simple indole alkaloids which incorporate in their skeletons a C_2 sub-unit derived from pyruvate. The mechanism of formation, via a Schiff base, and subsequent nucleophilic attack, is surely analogous to that observed in the formation of isoquinoline alkaloids. Tracer experiments with labelled pyruvate have substantiated this biogenetic hypothesis (eqn 5.25). Harmine and similar

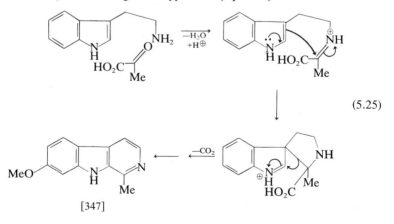

(5.25)

[347]

co-occurring alkaloids are believed to be the psychoactive principles of various South American 'magic' potions and snuffs.

The incipient nucleophilicity at C-3 can be invoked to explain how physostigmine [348], from the calabar bean, is derived from tryptophan. A concerted ring closure and nucleophilic attack on S-adenosyl methionine appears reasonable (eqn 5.26).

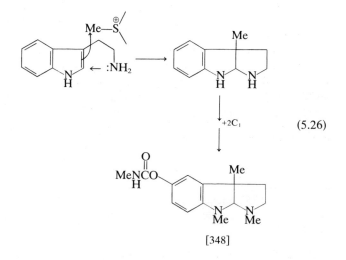

$$(5.26)$$

[348]

This mechanism must, however, await experimental proof, and the source of the methylcarbamate functionality is unknown. Physostigmine, and other co-occurring metabolites of similar structure, are very poisonous and were at one time used in trials by ordeal. Extracts of calabar beans were given to the defendant, and survival implied innocence, while death implied guilt. If the brew was consumed rapidly—and an innocent person might well do this since he or she would not fear the test—it caused vomiting and the physostigmine was expelled. A guilty person, on the contrary, might well sip the brew, fearing the consequences, hence allowing the alkaloid to be absorbed via the gastrointestinal tract and thus causing death. The test, although somewhat unreliable, might have been accurate in a good proportion of cases.

Alkaloids derived from anthranilic acid

Several species of higher plants produce alkaloids which are derived from 3-hydroxyanthranilic acid [294], itself a metabolite of tryptophan. Thus, damascenine [349] appears to be derived from 3-methoxyanthranilic acid [350], as shown in eqn (5.27).

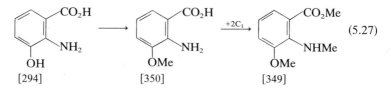

(5.27)

By far the largest number of these alkaloids occur in the family Rutaceae, and have structures based upon the quinoline and acridine systems. These are believed to derive from anthranilic acid with addition of C_2 units from malonate, and representative examples are shown in Fig. 5.24. Such structures may often be further modified by incorporation of a C_5 unit from MVA, and although these particular metabolites will be considered in more detail in Chapter 6, a few examples are given in Fig. 5.24. Feeding experiments with a cell suspension culture of *Ruta graveolens* have shown that $[1,2-^{13}C]$-acetate gives two superimposable labelling patterns for rutacridone, and this must be due to rotation of a symmetrical intermediate, prior to formation of the acridone ring system.

Finally, the 3-hydroxyanthranilic acid skeleton is clearly discernible in the structures of the fungal pigment cinnabarinic acid [351], and of the actinomycins [352] found in strains of *Streptomycetes*. Actinomycin A was isolated soon after penicillin, but though it possessed useful antibiotic

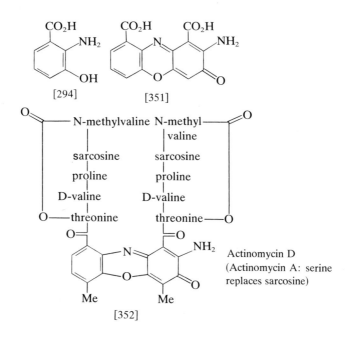

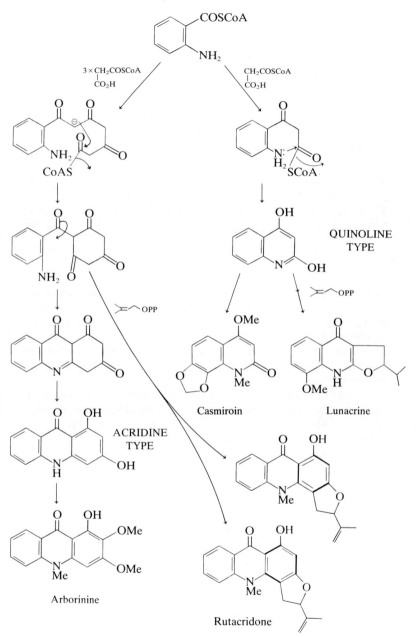

Casmiroin

Lunacrine

QUINOLINE
TYPE

ACRIDINE
TYPE

Arborinine

Rutacridone

Fig. 5.24

properties, its toxicity precluded clinical use. The actinomycins were 'rediscovered' when the compounds were evaluated for cancer chemotherapy about twenty years later. Actinomycin D rapidly and completely blocks synthesis of RNA by transcription from DNA, and since cancer cells usually divide (and need to synthesize RNA) at a faster rate than non-neoplastic cells, they are killed in greater numbers. It is thus one of the most potent antitumour agents known, though not selective enough for routine use.

Before leaving the alkaloids it is worth noting that less than 5 per cent of all known plant species have been screened for alkaloids. In light of their interesting pharmacological properties, it is hardly surprising that many pharmaceutical companies have initiated screening programmes on plants collected from less accessible or developed parts of the world. Those plants which play a part in native folklore and medicine are of particular interest, and there is obviously much still to be learnt about these fascinating metabolites.

Other metabolites derived from amino acids

In bacteria and other lower organisms where the metabolic pathways leading to alkaloids are generally absent, a large number of compounds are nonetheless formed from the metabolism of amino acids. These fall into two main structural classes: simple (though apparently non-essential) amino acids, and derivatives of peptides. Few biosynthetic experiments have been conducted, and the functions of these metabolites are unknown.

Among the metabolites produced by bacteria and fungi we have already mentioned chloramphenicol [251], which is a broad spectrum antibiotic, and now produced by chemical synthesis. It appears to inhibit the growth of bacteria by interfering with bacterial protein synthesis.

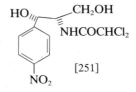

[251]

Cycloserine (oxamycin) [353] is produced by an actinomycete, and although it has much the same effects as penicillin on growing bacterial cells (see later), it is too toxic for clinical use. It is structurally similar to D-alanine [354], and appears to act as an anti-metabolite of this amino acid: it almost certainly inhibits the enzyme that epimerizes L-alanine to D-alanine, and as we shall see later when discussing the mode of action of the penicillins and cephalosporins, this enzyme is vital for the synthesis of bacterial cell walls.

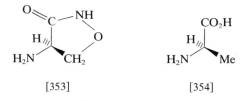

[353]   [354]

Certain higher plants also produce odd amino acids: the family Leguminosae produces over sixty toxic amino acids, and a few representative examples are shown in Fig. 5.25. β-Cyanoalanine is derived from the amino acid cysteine, and is toxic to mammals, probably because it inhibits the action of vitamin B_6. Mimosine, too, is toxic to a wide range of mammals, and is probably formed from the amino acid serine. As much as 8 per cent by weight is present in fresh leaves of one species of legume. Azetidine-2-carboxylic acid is the major free amino acid of lily of the valley, and also occurs in other plant species. It is the lower homologue of the amino acid proline, and its toxicity is probably due to interference with proline biosynthesis or utilizaton. It is derived from 2,4-diaminobutyric acid. A West African legume contains 5-hdroxytryptophan, and is credited with antiseptic and aphrodisiac properties. In the mammalian brain the metabolite is converted into 5-hydroxytryptamine, serotonin [342], which is psychoactive. Coprine, produced by the inky cap mushroom has antabuse activity, i.e. it mimics the activity of the drug used to treat chronic alcoholism. Both compounds inhibit alcohol dehydrogenase. The compound hypoglycin A is found together with the lower homologue—α-methylene, cyclopropyl-glycine—in the unripe fruit of the plant *Blighia sapida*, producing a dramatic lowering of blood sugar if the fruit are eaten. The effect is so dramatic (death may result!) that the compounds seem likely to have a role as feeding deterrents for herbivores. The utility of these compounds to the plants concerned is unknown, though a purely protective role is unlikely.

Another amino acid of interest is S-(*trans*-propen-1-yl)-cysteine sulphoxide [355], which is the precursor of the lachrymatory factor of onions, propane-thial-S-oxide [356] (eqn 5.28). Similar compounds are also found in garlic, and the complex sequence of reactions by which these metabolites are prepared is also shown in Fig. 5.25. The compound

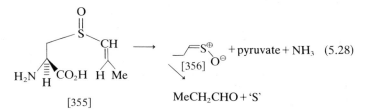

$$\text{[355]} \longrightarrow \text{[356]} + \text{pyruvate} + NH_3 \quad (5.28)$$

$$MeCH_2CHO + \text{'S'}$$

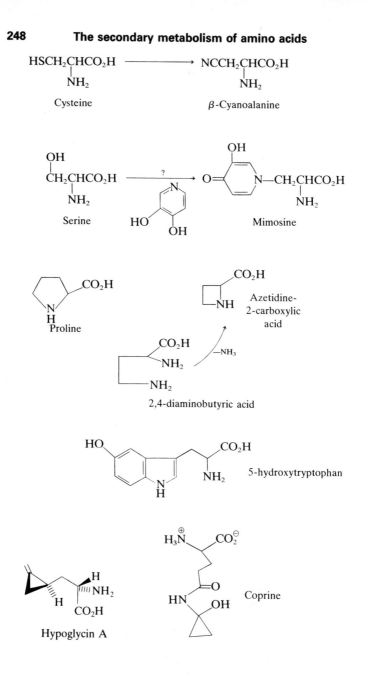

Fig. 5.25

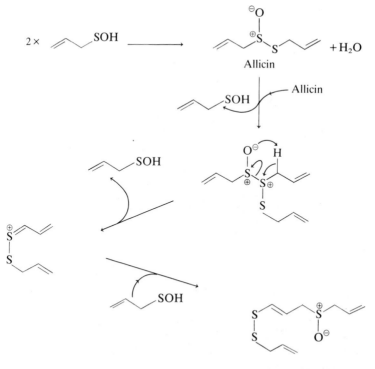

Fig. 5.25 *(continued)*

ajoene (a mixture of *E* and *Z* isomers) is of particular interest since it has been reported that it is a very potent inhibitor of aggregation of blood platelets. This may help to explain the supposed anti-thrombotic effect of garlic, and why the incidence of coronary thrombosis and stroke is lower in France (high consumption of garlic) than in the U.K. (low consumption).

The cyanogenic glycosides should be mentioned briefly at this point. These compounds, such as linamarin [357] and dhurrin [358] occur widely, and release hydrogen cyanide on hydrolysis, thus acting as effective feeding deterrents. They are probably derived via the sequence shown in eqn (5.29). In addition to their role as feeding deterrents, several cyanogenic glycosides have been shown to be potent allelopathic agents, and it is certain that they play a vital part in enhancing the competitiveness of those plants which produce them.

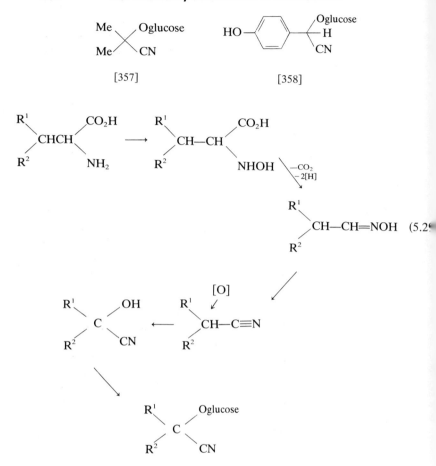

Finally, before considering peptide derivatives, we should consider amino acid polymers, which occur widely as pigments: for example melanins of animals, insects, and plants. Strictly speaking these are primary metabolites since they are indispensable to the organism. Thus the absence of a capability to produce melanin in skin, hair, and the retina characterizes the syndrome of albinism. Only one enzyme is concerned in the conversion of tyrosine to the polymer melanin, and this is tyrosinase which contains a copper ion as cofactor. Conversion of an O-quinone into polymers of various sizes is apparently spontaneous. In insects, at pupation, a metabolic pathway of tyrosine is altered, and the amino acid is diverted into formation of a melanin, which forms the basis of the pupal coat.

Since these polymers retain functionality (e.g. carbonyl groups, and —NH groups) they may interact with proteins by forming hydrogen bonds to the functional groups present in these molecules (—SH, —NH$_2$, —OH, —CO$_2$H), and in consequence melanins are not only pigments, but may also have a structural role.

Peptide derivatives

The peptides produced by bacteria and fungi are, with few exceptions, antibiotics. Amongst these we may distinguish at least two main groups: the cyclic peptide antibiotics, and the β-lactam antibiotics including penicillins, cephalosporins, thienamycins, newly discovered norcardicins and monobactams. Some typical metabolites are shown in Fig. 5.26.

Although the biogenesis of these compounds has been rather neglected, it is clear that the peptide link (—CONH—) is not formed in the same way as it is in protein biosynthesis, where messenger RNA (m-RNA) provides the nucleic acid 'code' which is translated into a polypeptide sequence through the agency of various transfer RNA (t-RNA) species. These carry an amino acid and a trinucleotide 'code-word' specific for that amino acid. Matching of the m-RNA trinucleotide 'codon' and the t-RNA 'anti-codon' results in incorporation of the amino acid on the growing polypeptide chain. This process is highly specific: only one, unique polypeptide can be synthesized through the agency of a particular m-RNA.

In contrast the process involved in the synthesis of peptide derivatives frequently results in the production of a series of analogues, for example the actinomycins [352] discussed previously, or the penicillins, where a number of different amino acid side-chains are encountered. One common (though not universal) feature, however, is the activation of the amino acids by conversion into the amino acyl adenylates prior to incorporation into the growing polypeptide chain (eqn 5.30). The

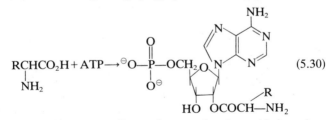

incorporation of the 'unnatural' enantiomers of amino acids into these peptides is also commonplace and D-amino acids probably arise entirely by racemization of L-amino acids.

Some progress has recently been made concerning the biogenesis of the penicillins and cephalosporins. The natural occurrence of cephalosporin C [359] and isopenicillin N [360], which have the same side-chain but

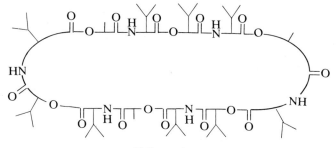

Valinomycin

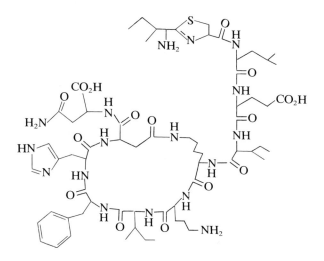

Bacitracin A

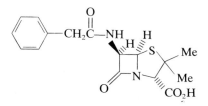

Benzyl penicillin (penicillin G)

Fig. 5.26

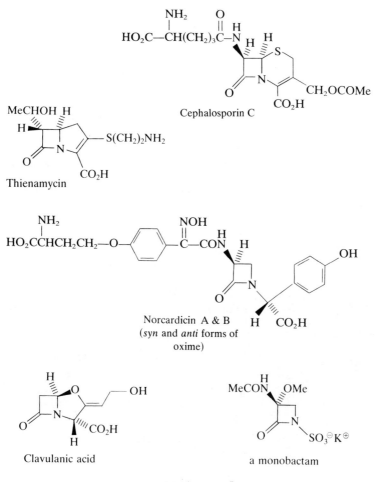

Fig. 5.26 (*continued*)

different nuclei, suggest a common biosynthetic pathway via δ-(L-α-aminoadipoyl)-L-cysteinyl-D-valine [361] as shown in Fig. 5.27. This compound has been isolated in small amounts from cultures synthesizing penicillins. Tracer and enzyme studies have demonstrated that the three constituent amino acids are incorporated into [359] and [360], and have also given some indication of how the five-membered (penam) ring and six-membered (cepham) ring may arise. These experiments are shown in diagrammatic form in Figs 5.27 and 5.28. There is some evidence that β-lactam ring formation occurs prior to construction of the thiazolidine

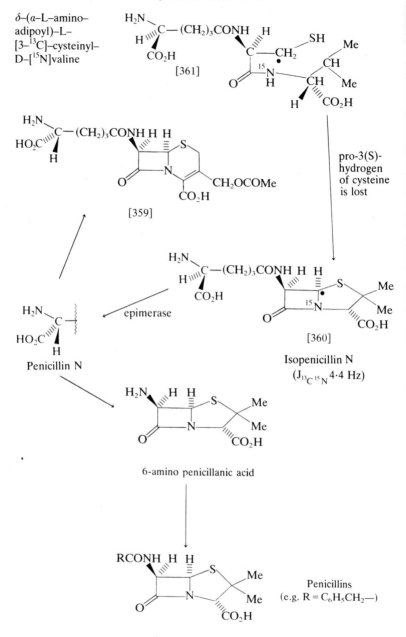

Fig. 5.27

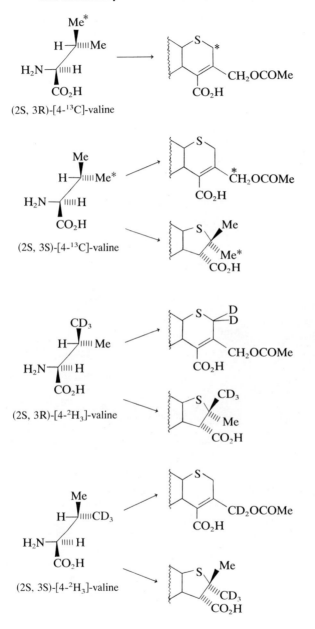

Fig. 5.28

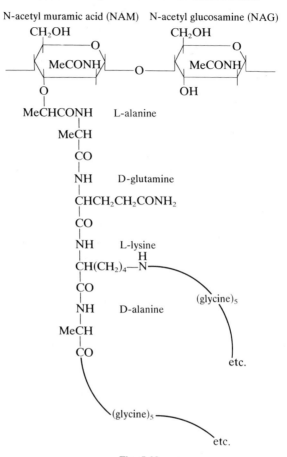

Fig. 5.29

ring of the penicillins, but to date no intermediates between the tripeptide [361] and isopenicillin N [360] have been isolated.

As to the mode of action of these antibiotics, it is now quite clear that they interfere with the biosynthesis of the bacterial cell wall. They are most active against Gram-positive organisms (Streptococci, Staphylococci, etc.), and only the newer synthetic penicillins and cephalosporins are active against Gram-negative bacteria (*E. coli*, Salmonella, etc.). This difference in reactivity presumably reflects the differing modes of bacterial cell wall assembly.

The cell wall is composed of repeating disaccharide-tetrapeptide units, which are cross-linked in a three-dimensional matrix. The sugars and amino acids utilized differ from species to species, but the structure of the cell wall of *Staphylococcus aureus* is exemplary (Fig. 5.29). In this case the cross-links are provided by a pentaglycine bridge between the 6-amino group of lysine of one chain, and the terminal D-alanine of another chain. Prior to formation of these cross-links, the sub-unit at the cell periphery is [362] (see also Fig. 5.29).

NAG—NAM—phospholipid (glycine)$_5$ [362]

L-alanine—D-glutamine—L-lysine—D-alanine—D-alanine

An extra D-alanine residue is present, and penicillins and cephalosporins prevent cross-linking (i.e. polymerization) by inhibiting the transpeptidase that catalyses removal of this extra alanine: subsequent cross-linking via pentaglycine bridges is probably spontaneous, and the phospholipid is also cleaved. The antibiotics are believed to act as irreversible enzyme inhibitors, and there is sufficient structural similarity between D-alanyl-D-alanine [363] and the common sub-structure of these compounds [364] such that the latter might be accepted at the enzyme active site. The lactam ring is cleaved and irreversible acylation of a vital functional group (i.e. —OH to —OCOR or —NH$_2$ to —NHCOR) at the active site is thought to be the mode of inhibition.

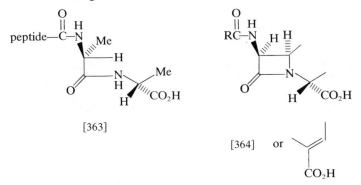

[363]

[364]   or

In the light of the preceding comments, it is notable that these antibiotics only kill growing cells, that is cells which need to synthesize new cell-wall material. In effect the cell grows out of its protective coat, and the sensitive inner components are irreversibly damaged. One problem with all of these antibiotics is the emergence of resistant bacteria which possess β-lactamases. These enzymes cleave the β-lactam ring thus rendering the antibiotics inactive, and an extensive search has been carried out by the pharmaceutical industry for β-lactamase inhibitors. The compound clavulanic acid (see Fig. 5.26) from *Streptomyces clavuligerus* is just such a compound. It is a poor antibacterial substance but is a potent β-lactamase inhibitor, and is administered in conjuction with a penicillin in order to protect and thus potentiate the activity of the antibiotic. Synthetic analogues of the monobactams are also of current interest since they combine useful antibacterial activity with resistance to β-lactamase inactivation.

The significance of these peptide derivatives to the organisms producing them is still completely unknown. A beneficial role is hard to reconcile when we note that only a small proportion of all of the known soil microorganisms produce peptide derivatives. In addition, the quantities of these compounds produced under natural conditions are really rather small. It is possible that the cyclic peptides mediate the transfer of cations across biological membranes, and this play a vital part in controlling membrane permeability. They certainly are avid sequestors of cations, notably of Na^+, K^+, and Ca^{2+} ions, and in this respect they resemble the so-called crown ethers, such as [365]. (Both the cyclic peptides and the

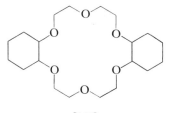

[365]

crown ethers probably function as giant chelating species.) However, although one or two proven examples of membrane transport have been reported (e.g. bacitracin participates in transport of Mn^{2+} across the cell membranes of *Bacillus licheniformis*), a definitive explanation of the mode of action of the cyclic peptides is awaited with interest.

Problems

5.1. You have just characterized pinidine. How would you attempt to elucidate its biosynthetic pathway?

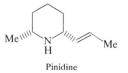

Pinidine

[Leete, E., Lechleiter, J. C., and Carver, R. A. (1975). *Tetrahedron Lett.*, 3779]

5.2. (difficult). The alkaloid chelidonine was labelled as shown when [2-^{14}C]-tyrosine was employed as precursor. Account for this result: stylopine was an isolated intermediate.

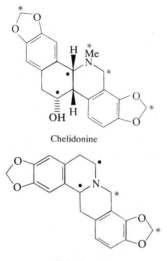

Chelidonine

Stylopine

[2-^{14}C]-tyrosine (•)
from C$_1$ pool (∗)

[Battersby, A. R. Francis, R. J., Southgate, R., Staunton, J., and Wiltshire, H. R. (1975). *Perkin Trans. 1*, 1147]

5.3. (difficult). The antibiotic anthramycin ex. *Streptomyces refuineus* was labelled (as shown) when [^{13}C]-methionine (∗) and [1-^{13}C]-tyrosine (•) were employed as precursors. The aromatic ring is known to derive from tryptophan (probably via 3-hydroxyanthranilic acid). Suggest a plausible biosynthetic pathway.

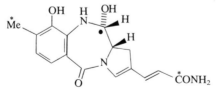

[Chang, C., Hurley, L. H., and Zmijewski, M. (1975). *J. Am. Chem. Soc.*, **97**, 4372]

5.4. The alkaloid vertine is produced by the plant *Heimia salicifolia*. If [4,5-^{13}C]-lysine was used in a feeding experiment, which carbon atoms would exhibit enhanced signal? Also suggest a precursor of the other parts of the molecule.

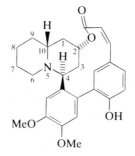

[Hedges, S. H., Herbert, R. B., and Wormald, P. C. (1983). *Chem. Commun.*, 145]

5.5. The mould *Penicillium cyclopium* produces the benzodiazepine cyclopenin. Identify the probable amino acid precursors of this metabolite.

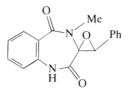

[Kousy, S. E., Luckner, M, Nover, L., Schwelle, N., and Voigt, S. (1978). *Phytochemistry*, **17**, 1705]

6. Metabolites of mixed biosynthetic origin

We have already encountered a few examples of compounds which have a mixed biosynthetic origin: that is, they incorporate within their structures the biogenetic sub-units of two (or more) metabolic pathways. However, these examples have been restricted to those metabolites which are derived primarily via one metabolic pathway, with minor skeletal modifications due to the introduction of one C_5 sub-unit (mevalonate pathway) or of one (and sometimes more) C_2 unit (polyketide pathway). The compounds to be considered in this chapter have skeletons which are derived from approximately equal contributions of sub-units from at least two metabolic pathways. Most of these compounds are comparatively rare, occurring in only one species or in a number of closely related species; but others, such as the flavonoids and polyisoprenyl quinones, are of widespread occurrence and vital importance.

Metabolites derived from acetate and mevalonate

A classical example of a compound of mixed biogenetic origin is provided by the mould metabolite mycophenolic acid [366]. Birch established that alternate carbon atoms were labelled when [1-^{14}C]-acetate was fed to *Penicillium brevicompactum*, and thus provided the first compelling evidence in support of the 'acetate hypothesis'. Subsequently, the mevalonoid origin of the side-chain was also demonstrated (eqn 6.1).

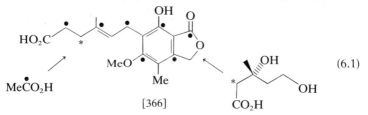

(6.1)

[366]

Birch and his coworkers originally believed (and there was no reason at the time to suppose that it was otherwise) that a C_{10} geranyl unit was incorporated, and then partially degraded. Recent experiments have, however, demonstrated that a C_{15} moiety is utilized, presumably farnesyl pyrophosphate [122], and the presently accepted mode of biosynthesis is shown in Fig. 6.1. The points at which introduction of the two methyl groups and the farnesyl side-chain occur are known from experiments in which radio-labelled intermediates [367]–[369] were isolated, and then

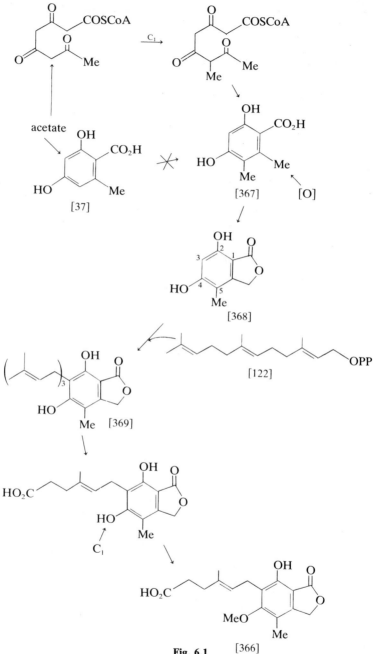

Fig. 6.1

incorporated into mycophenolic acid. Thus, labelled orsellinic acid [37] was not incorporated but intermediate [367] (2,4-dihydroxy-5,6-dimethylbenzoic acid) was incorporated very efficiently. Methylation must then occur prior to ring closure. Similarly intermediate [368] was efficiently utilized by the fungus, but its 4-O-methyl ether was not.

Farnesyl pyrophosphate is also a precursor of the siccanochromenes, e.g. siccanin [370], which are produced by the fungus *Helminthosporium siccans*. Again a polyketide origin has been established for the aromatic portion of the molecule (Fig. 6.2).

The cannabinoids
Preparations from the plant *Cannabis sativa* have been used for thousands of years, but have achieved a quite unparalleled notoriety in recent times. An estimated 200–300 million people use marihuana, hashish, bhang, charas, daga, etc.; these are names which are associated with the crudity of the preparation or identify the part of the plant from which it derives. Although the major psychoactive component, $(-)$-Δ^1-tetrahydro-

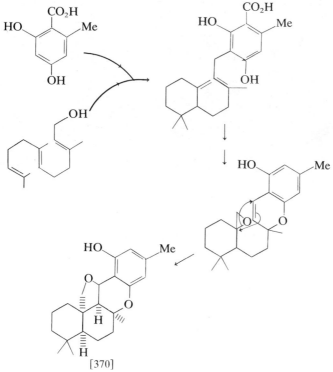

[370]

Fig. 6.2

cannabinol (Δ^1-THC) [317], is known, little reproducible biological work has been carried out, and our knowledge of the pharmacology of marihuana is limited. Nonetheless, marihuana is outlawed in all

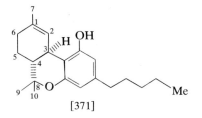

[371]

Western countries, and generally classified with narcotics such as morphine and other abused psychoactive agents like cocaine and LSD. Presently available evidence seems to refute this assignment, and suggests that it may be more accurate to classify it with alcohol as an agent which may cause physical harm, and may be abused.

Δ^1-THC is the major naturally occurring cannabinoid, and varying amounts of $\Delta^{1(6)}$-THC [372] (also psychoactive), cannabidiol [373] (R = H), cannabinolic acid [374] (R = CO_2H) co-occur in most plants of the species.

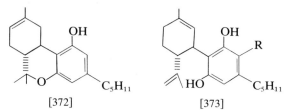

[372]   [373]

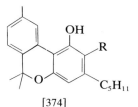

[374]

Labelling studies have been rather unsuccessful due to poor incorporations of labelled substrates, but it is generally accepted that the cannabinoids are derived from geranyl pyrophosphate [103] and olivetol [375], which is derived from a polyketide precursor. Possible biosynthetic intermediates like cannabigerol [376] and cannabidiol [373] (R = H) have been suggested, and the pathway shown in eqn (6.2) has been

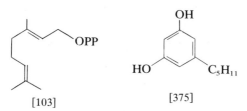

[103]   [375]

proposed. Both [375] and [376] have been incorporated into [371] during labelling studies. However, cannabidiol is not present in some samples of *Cannabis sativa* or hashish, and alternative intermediates may

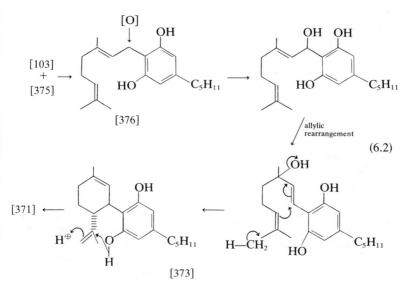

(6.2)

be utilized, e.g. [377]. In the light of a recent, direct (biomimetic?) synthesis of Δ¹-THC (eqn 6.3), even intermediates like [378] are conceivable. In short, we remain largely ignorant of the biosynthetic pathway to the cannabinoids.

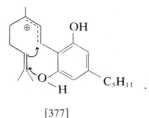

[377]

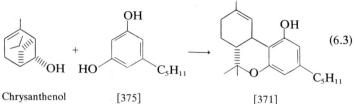

Chrysanthenol   [375]   [371]   (6.3)

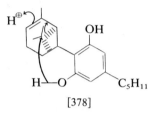

[378]

Metabolites derived from shikimate and mevalonate

Isoprenoid quinones

The isoprenoid quinones: ubiquinones, plastoquinones, tocopherols, and menaquinones, comprise the largest group of metabolites which derive from shikimate and mevalonate, and are not strictly speaking secondary metabolites, since they are ubiquitous, and have essential biological roles in most instances. A complete understanding of the various biosynthetic pathways is still lacking, but sufficient tracer results have been obtained to allow reasonable proposals to be made.

The ubiquinones, which are almost without exception 5,6-dimethoxy-3-methyl-2-all-*trans*-polyprenyl-1,4-benzoquinones   [379],

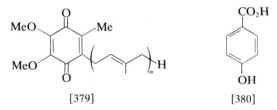

[379]   [380]

have been detected in most living organisms. The two major exceptions are Gram-positive bacteria and blue-green algae. They are vital components of respiratory chain phosphorylation and bacterial photophosphorylation, i.e. processes whereby electron transfer via a series of redox systems is coupled to production of ATP. The most widespread ubiquinones are those in which there are 8, 9, or 10 isoprene units, [379] ($n = 8$, 9, *and* 10). In addition a few ubiquinones possess modified side-chains (e.g. monoepoxide, saturated terminal isoprenoid unit). In animals phenylalanine and tyrosine act as precursors of 4-hydroxybenzoic acid, [380], the last common intermediate prior to incorporation of an intact polyisoprenoid side-chain. Bacteria utilize chorismate, [216], as precursor. The sequence of reactions thereafter has been delineated through extensve use of mutants of the bacterium *Escherichia coli*: each 'useful' mutant lacks the capacity to carry out one of the transformations, and the substrate of this reaction thus accumulates and may be isolated and characterized. The pathway of ubiquinone biosynthesis, at least for *E. coli*, is shown in Fig. 6.3: both methoxyl oxygen atoms are derived from molecular oxygen, and the three methyl groups from S-adenosyl methionine. By contrast, in eukaryotes (organisms whose cells contain true nuclei) hydroxylation and methylation precede the decarboxylation step. In fact O-methylation appears to be of regulatory significance since yeast (*Saccharomyces cerevisiae*) accumulated dihydroxy-species when biosynthesis of ubiquinone-6 ($n = 6$) was depressed by addition of glucose.

The plastoquinones and tocopherols are biosynthesized almost exclusively in higher plants and in algae, and differ from the ubiquinones in having a variety of possible side-chain modifications. The basic structures are shown in Fig. 6.4. Plastoquinones participate in electron transfer during the light reaction of photosynthesis, and are found primarily in the chloroplasts of plants. α-Tocopherol is the main active principle of vitamin E (β- and γ-tocopherol are also present but are less potent). This is the 'wheat germ' vitamin, and appears to have a role in electron transport, and perhaps as an antioxidant protecting unsaturated fatty acids from autoxidation. Vitamin deficiency is associated with muscular dystrophy in animals, but not, apparently, in Man. At present vitamin E preparations are very popular in the U.S.A. due to an alleged association with maintenence of sexual potency: there is little evidence to support this supposition.

Biosynthetic details have only recently become available and the scheme shown in Fig. 6.5 has been proposed. It is possible that isoprenylation and decarboxylation of homogentisate occur concomitantly, but this remains to be established, and the remaining steps of the pathways are unknown. It is, however quite probable that the tocopherols are derived as shown in equation (6.4). The most noteworthy feature is that in contrast to the biosynthesis of the ubiquinones, where only the aryl unit of *p*-hydroxybenzoic acid [380] is retained in the final metabolite, these metabolites retain an ArC_1 unit of the intermediate (homogentisate) which is the substrate for isoprenylation.

The final members of the family of isoprenoid quinones are the phylloquinones [381] and menaquinones [382]. The former are only

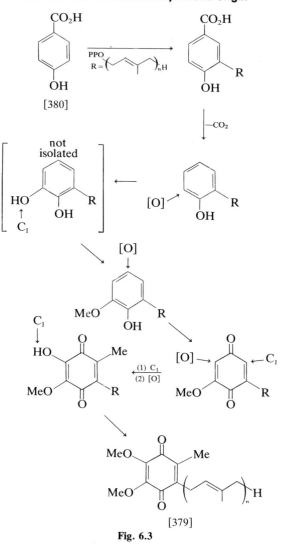

Fig. 6.3

produced by higher plants and algae, and the latter by bacteria and fungi. The phylloquinone which bears the methyl group is vitamin K_1, and menaquinone-6 (i.e. $n = 6$ or 7) is vitamin K_2. They are required to promote formation of prothrombin and other plasma proteins which are essential for normal coagulation of blood (i.e. clotting). Any vitamin deficiency caused either by dietary insufficiency, or through deficient synthesis by intestinal flora, leads to an increased clotting time, and

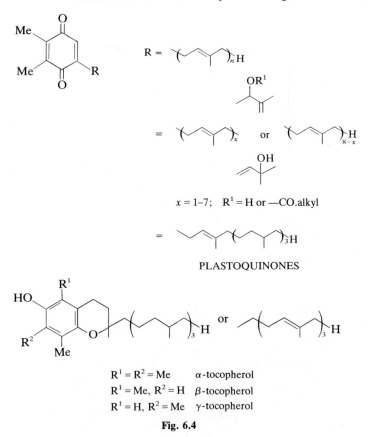

PLASTOQUINONES

$R^1 = R^2 = Me$ α-tocopherol
$R^1 = Me, R^2 = H$ β-tocopherol
$R^1 = H, R^2 = Me$ γ-tocopherol

Fig. 6.4

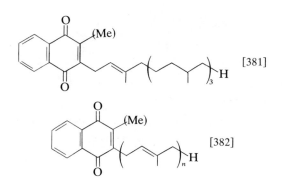

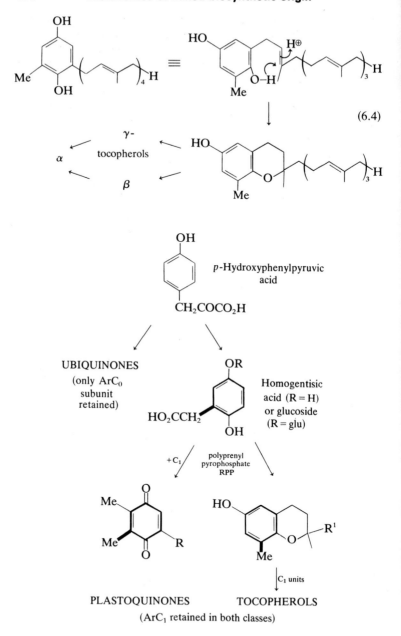

(6.4)

Fig. 6.5

perhaps to haemorrhaging. The rodenticide warfarin [224] interferes with the normal clotting process; and vitamin K and warfarin are mutually antagonistic, perhaps competing for the same enzyme active site.

The biosynthetic routes to these metabolites are still unknown, but some recent experiments have established that a symmetrical intermediate is not involved. A pathway utilizing shikimate and 2-ketoglutarate [383] is envisaged (Fig. 6.6), and the symmetrical intermediate, [384], is not involved.

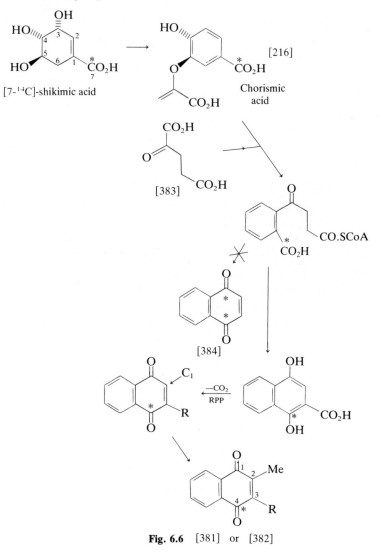

Fig. 6.6   [381]   or   [382]

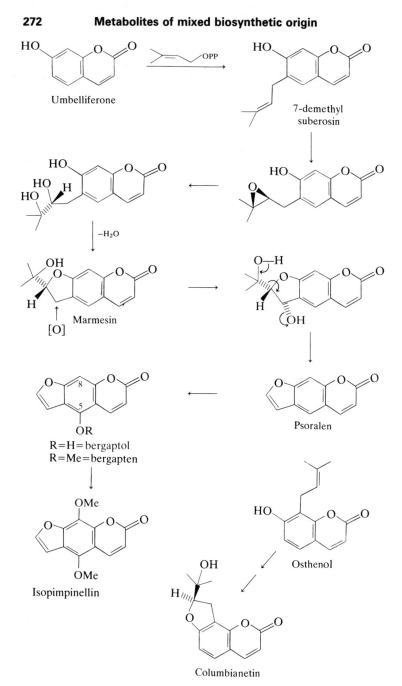

Umbelliferone

7-demethyl
suberosin

Marmesin

[O]

Psoralen

R=H=bergaptol
R=Me=bergapten

Isopimpinellin

Osthenol

Columbianetin

Fig. 6.7

Furanocoumarins and furanoquinolines

The furanocoumarins such as marmesin and columbianetin (see Fig. 6.7) contain a coumarin sub-unit and a C_5 sub-unit. Biosynthetic pathways via 7-demethylsuberosin and osthenol (respectively) have been delineated, and the route to marmesin is shown in Fig. 6.7. Presumably the pathway by which columbianetin is derived is analogous. Further metabolism of marmesin produces psoralen, in which three of the original carbon atoms have been lost. Hydroxylation (at C-5) then yields bergaptol and methylation yields bergapten; subsequent hydroxylation (at C-8) and methylation finally produces isopimpinellin. An alternative pathway is

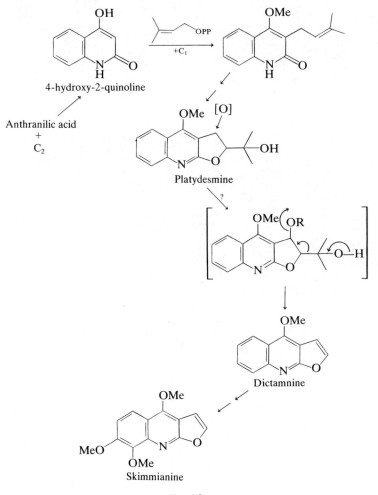

Fig. 6.8

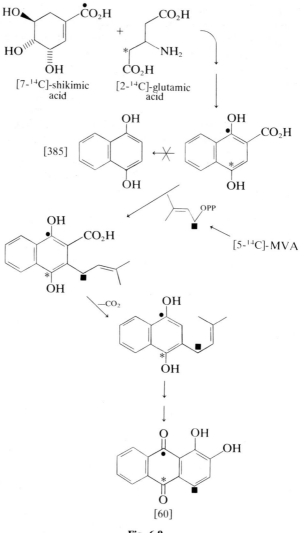

Fig. 6.9

also present in several plants, and this proceeds via hydroxlation at C-8 followed by methylation and hydroxylation at C-5. Not surprisingly, two separate O-methyltransferase enzymes are involved, and these have been isolated and characterized. The dimethylallyl transferase has also been isolated, and appears to be located mainly in the chloroplasts. As mentioned previously, it was the co-occurrence of marmesin and psoralen which provided the clue that led to identification of a common metabolic pathway. It is, after all, not immediately obvious that psoralen is derived from a shikimate sub-unit and a C_5 sub-unit, since it does not possess the requisite number of carbon atoms. The furanoquinolines were briefly mention in Chapter 5, and their presumed mode of biosynthesis is closely similar to that just discussed (Fig. 6.8).

Alizarin
It has already been noted (in Chapter 2) that many anthraquinones are derived via the polyketide pathway. Several metabolites of this type are however derived from shikimate and MVA. Thus labelling studies suggest that alizarin [60] is biosynthesized as shown in Fig. 6.9. Once again a symmetrical intermediate, e.g. [385], is not involved, i.e. decarboxylation must occur after introduction of the C_5 sub-unit.

For centuries alizarin, as extracted from the madder plant, has been used to impart a purplish red colour to fabric. Recently another antiquarian use has been unearthed: at Qumran, which has been associated with the Dead Sea Scrolls, the inhabitants were in the habit of consuming the madder plant as a protection against illness and evil spirits. Since alizarin has an affinity for those parts of the skeleton where *de novo* formation of bone occurs (it probably forms a calcium salt), the bones of these people, excavated near Qumran, are stained purplish red. To this day certain Arab tribes believe that madder has magical properties, and a sherbet drink containing madder extract is taken as a defence against the 'evil eye'.

Metabolites derived from acetate and shikimate

Flavonoids
The flavonoids comprise a large group of secondary metabolites which are derived from sub-units supplied by the acetate and shikimate pathways. They occur almost exclusively (usually as glycosides) in higher plants and are responsible for much of the flavour of food and drink of plant origin and for the colour of flowers. A basic C_{15} unit is invariably present, and this is shown as structure [386]. It is well established from tracer experiments that the ArC_3 sub-unit is derived from shikimate, and that the other aromatic ring (ring A) is of polyketide origin. The usual oxygenation pattern of ring A is shown in [386], and ring B may have a

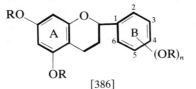

[386]

4-hydroxy, 3,4-dihydroxy, or 3,4,5-trihydroxy substitution pattern. Further oxidation, reduction, and alkylation of the basic skeleton may also occur to produce additional structural features.

The major flavonoid classes are shown in Fig. 6.10, together with their probable biogenetic derivations. Phenylalanine [210] (not tyrosine) is the precursor of the *p*-coumaryl sub-unit, which condenses with three C_2 units to produce chalcones. The hydroxylation pattern of the A ring is almost certainly established at the cyclization stage, while that of the B ring may not be determined until quite late in the sequence. Tracer and enzyme studies have established the intermediacy of chalcones, and the newly discovered α-hydroxychalcones may be key precursors in the biosynthesis of flavanonols (dihydroflavonols), flavonols, catechins, and anthocyanidins. In particular, extensive studies on the enzymes isolated from suspension cultures of parsley, have demonstrated the marked specificities of the key enzymes *p*-coumarate-CoA ligase and chalcone synthase. The former probably represents a control point for the utilization of *p*-coumaryl-CoA by the various phenylpropanoid pathways (see also Chapter 4). Some chalcone synthases, in contrast, will accept *p*-coumaryl-SCoA and caffeyl-SCoA, but there is a different pH optimum for each substrate.

Most of the skeletal alterations may be explained in quite simple chemical terms, although little definite information is available. In particular the formation of isoflavonoids is intriguing. An isoflavone synthase has been isolated, and appears to function as a monooxygenase. The proposed pathway is shown in eqn. (6.5).

We shall now consider some representative flavonoids. *Dihydrochalcones* are comparatively rare, but a well-known example is phlorizin [387], a component of the root bark of apple trees, which

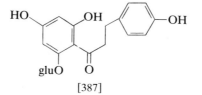

[387]

confers disease resistance on the apple plant. On ingestion this compound apparently impairs reabsorption of glucose by the kidneys, and thus produces a diabetes-like condition (i.e. excess sugar in blood and urine). It is also used by the motor trade as an additive in motor oil!

Chalcones are also quite rare, due at least in part to their ready isomerization to flavanones, but one that is of some interest is a yellow pigment in the petals of safflower [388]. As the flower ages a red pigment carthamin, is produced, and to this has been assigned structure [389].

The *flavanones* are also relatively uncommon, but occasionally

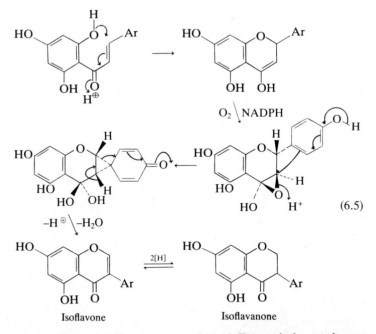

$$(6.5)$$

Isoflavone   Isoflavanone

accumulate in fruits, flowers, leaves, and wood. Two typical examples are soluble, bitter-tasting compounds from citrus, e.g. naringin (naringenin glycoside) [390] from grapefruit peel, and hesperidin (hesperetin glycoside) [391] from orange peel. It is interesting to compare the intensely bitter natural naringin with the very sweet-tasting synthetic naringin dihydrochalcone [392], since the two structures are rather similar. Whether these bitter compounds, and others like them, are mediators of plant–insect interactions (i.e. as repellents) is a matter of conjecture, and little experimental work has been carried out.

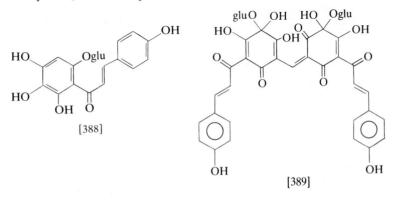

[388]

[389]

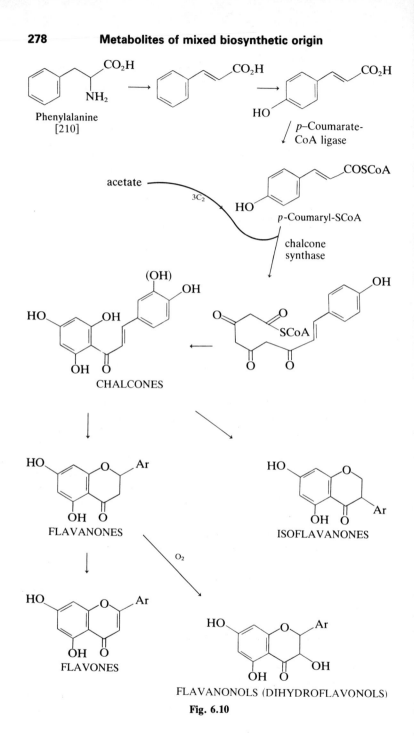

Fig. 6.10

FLAVANONOLS (DIHYDROFLAVONOLS)

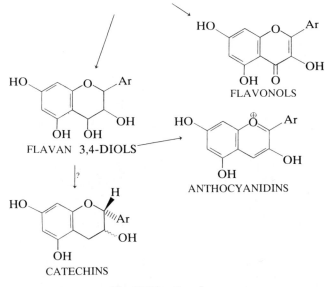

FLAVONOLS

FLAVAN 3,4-DIOLS

ANTHOCYANIDINS

CATECHINS

Fig. 6.10 (*continued*)

Both *flavones* and *flavonols* are very widely distributed, and luteolin (flavone) [393] and quercetin (flavonol) [394] are particularly common in leaves, but are also found as constituents of rinds and barks, in clover blossom, and in ragweed pollen.

The *isoflavonoids* occur most commonly in plants of the family Leguminosae, and as shown in eqn (6.5), it is believed that isoflavones are formed from flavanones (as their enol forms), and may then be converted into isoflavanones. Two interesting examples are provided by coumestrol [395] and rotenone [396]. The former is an oestrogenic agent (causes ovulation), found especially in clovers and in alfalfa and is presumably formed from the enol form of an isoflavanone as shown in equation (6.6). Rotenone is of interest as a natural insecticide and piscicide (fish poison), and was used for the latter purpose by primitive Man. The complete biosynthetic pathway to rotenone has been established by Crombie and coworkers.

The *catechins*, for example (+)-catechin [397], are primarily important as components of the proanthocyanidins. It has been suggested that these arise through reaction of a catechin with a quinone methide (eqn 6.7), and the latter is also a suggested progenitor of catechins.

Polymerization of flavan diols produces condensed tannins, of which structure [398] is a representative example. Tannins are often astringent, and their presence in plants undoubtedly has a protective effect, since

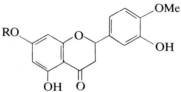

R = H = naringenin
R = 6-O-(α-L-
rhamnopyranosyl-O-glucopyranose)
= [390]

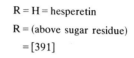

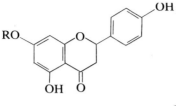

R = H = hesperetin
R = (above sugar residue)
= [391]

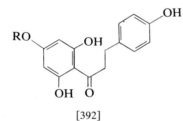

R = -glu-O-rham
(as in [390] and [391])

[392]

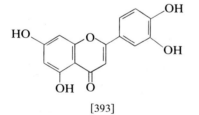

[393]

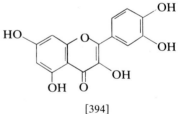

[394]

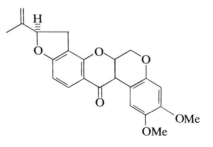

[396]

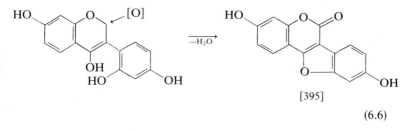

[395]

(6.6)

their taste is known to repel at least the larger herbivores. Other so-called hydrolysable tannins are derived from gallic acid and glucose, and a typical structure is shown as [399]. These tannins are also astringent and probably play a similar part in repelling herbivores.

The *anthocyanins*, glycosides of *anthocyanidins*, are primarily responsible for the red, blue, and violet colours of flowers and fruits. They are believed to derive from dihydroflavanols, but the final stages of the biosynthetic pathway remain to be elucidated. One of these steps, perhaps formation of the final cationic species, is light-controlled, though the significance of this fact is also unclear. Two common anthocyanidins (R = sugar residue in anthocyanins) are cyanidin [400] and malvidin [401] (both with R = H). The anthocyanidins are shown as cationic forms since flower sap is invariably acid; but *in vitro* solutions exhibit pH dependent colour changes: red at low pH (cation), through violet (neutral form), to blue (anion) at high pH.

The variegated colours of petals almost certainly act as attractive stimuli for pollinating insects. The anthocyanins thus have a vital role as mediators of these interactions. It is not surprising then to find that the variety of hues associated with the anthocyanins has been increased by evolutionary processes. Thus, natural selection (i.e. survival of the 'fittest') seems to have led to the production of scarlet hues, typical of pelargonidin [402], or blue hues, typical of delphinidin [403], through modification of the 'primitive' pigment, cyanidin [400], by loss or gain (respectively) of an hydroxyl group. Similarly, methylation leads to production of altered hues, probably due to subtle changes in the metal binding capabilities of the resultant compounds. Much recent enzymic work has been carried out to isolate and characterize the various O-methylases involved.

It should be noted that although flavonols and flavones are not coloured, they do absorb strongly in the UV, and although invisible to the human eye, can be seen by insects. They often occur at the centre of flowers, and probably act as 'honey guides': that is, they attract insects to their nectaries. In return the insects are accessories in the process of pollination, carrying away from the flower, not only nectar, but also pollen particles for transference to other plants.

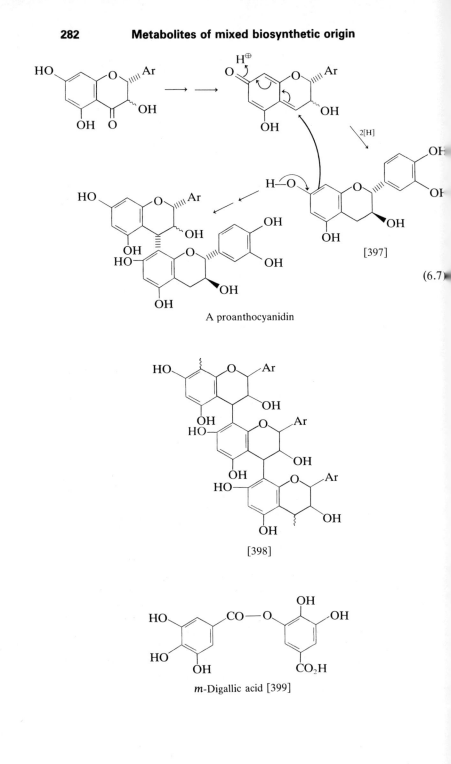

A proanthocyanidin

[397]

(6.7)

[398]

m-Digallic acid [399]

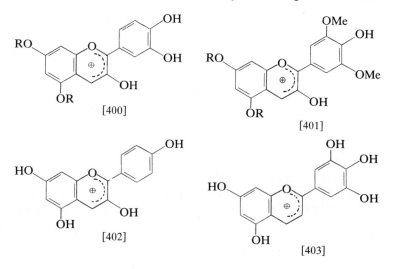

In passing, it is worth noting that the strong UV absorption of the flavonoids may be of more general utility, i.e. they may act as UV screens for endogenous nucleic acids (sensitive region 250–270 mm), and for NADH and other coenzymes (320–350 nm). The development of the metabolic pathways which lead to flavonoids would thus have been of benefit to plants that emerged from the primaeval oceans; and it is notable that marine plants do not produce flavonoids.

Some comment should be made concerning the life-times of these flavonoids, that is, their rate of turnover in plants. *In vivo* the flavonoids have quite long turnover times, with biological half-lives (i.e. the time taken for the initial number of moles of a compound to be reduced by half through metabolism) of several days in leaves and stems, but of the order of one day in petals (anthocyanins). The shorter life-times of the anthocyanins is hardly surprising in view of the shorter life-span of flowers when compared with the life-cycle of the plant.

However, once the *status quo* of the plant is disturbed, we encounter the now familiar situation of possible drastic structural alterations due to aerial oxidation or exposed to enzymes not normally encountered. Thus the compound isolated is not necessarily the one extant *in vivo*. Such *post mortem* changes are important in food technology, and the final flavour of chocolate, for example, appears at some point during fermentation and subsequent roasting of cocoa beans. This process is poorly understood, but almost certainly involves a complex series of changes in the flavonoids (mainly anthocyanins and flavonols) that are present in the beans. Similarly, the colour of red wine is due to polymerized anthocyanins formed from grape pigments during fermentation and storage.

The flavonoids have always been of interest to botanists and plant taxonomists. Since they occur in all land plants, unlike, for example, the alkaloids, which are much less widespread, they are widely used as taxonomic markers. In fact it has been amply demonstrated that the flavonoid content of a plant provides a clear indication of its evolutioary status (i.e. primitive or more recent). In many instances hybridization of two plants containing different flavonoids produces a new strain of plant (a hybrid) which is capable of synthesizing the flavonoids typical of both parents. This fact is of obvious importance to plant breeders because of the changes of colour (anthocyanins) produced in petals through hybridization.

Finally, many flavonoids are phytoallexins (i.e. produced by the plant in response to microbial invasion), and typical examples are pisatin [404] (ex. pea plants), and phaseollin [405] (ex. *Phaseolus vulgare*, the common

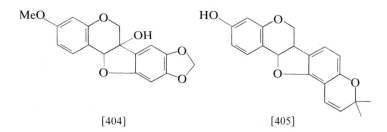

[404]   [405]

bean). Specific incorporation of [1,2-^{13}C]-acetate into pisatin has been demonstrated (eqn 6.8) using *Pisum sativum*; but this specificity is by no means general, and incorporation studies with other plants has shown that free rotation at the chalcone stage can lead to scrambling of the label.

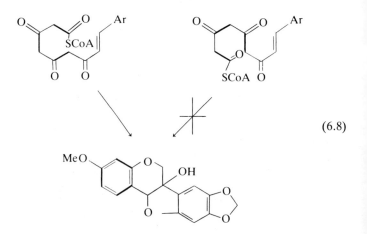

(6.8)

Xanthones, stilbenes, and related metabolites
Xanthones and stilbenes of plant origin are believed to be derived from shikimic acid (those of fungal origin are almost certainly derived solely from acetate see Chapter 2: Figs. 2.13 and 2.14), but few tracer studies have been conducted. From the results available probable pathways have been proposed for the biosynthesis of the xanthone, gentisein [406] (Fig. 6.11), and of the dihydrostilbene, lunularic acid [407] (Fig. 6.12). This latter compound is notable since it inhibits the growth of certain algae, and of liverworts (a group of plants related to the mosses). It appears to play the same role in these lower plants that abscisic acid [208] plays in higher plants.

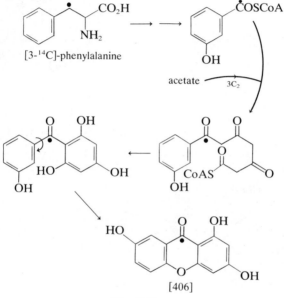

Fig. 6.11

Metabolites derived from tryptophan and mevalonate

As mentioned in Chapter 5, there are many alkaloids which incorporate within their structures one sub-unit derived from tryptophan [254] and one derived from mevalonate. There are two major classes: the vinca alkaloids and the ergot alkaloids; and in addition, several well-known compounds like quinine and strychnine. Although many of these metabolites are structurally very complex, a tryptamine unit (Indole C_2N) is usually discernible within the skeletons, and the remaining carbon atoms (usually an integral number of C_5 sub-units) are almost invariably derived from mevalonate. Some representative examples are ajmalicine

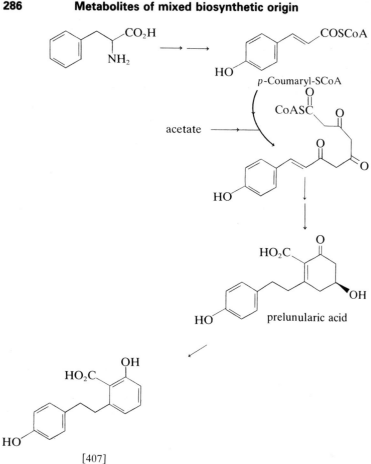

[407]

Fig. 6.12

[408], a vinca alkaloid, lysergic acid [101], an ergot alkaloid, strychnine [409], and quinine [410]. Many of these alkaloids have interesting pharmacological properties, and this has been an added incentive for their study.

Vinca alkaloids
Catharanthus roseus produces an abundance of exotic alkaloids, and the biosynthetic pathways by which they are produced have been exhaustively studied by Battersby, Arigoni, Scott, and others. Three basic structural types are encountered, typified by vindoline [411], catharanthine [412] and ajmalicine [408]. The realization, that the C_{10} sub-units (given in bold

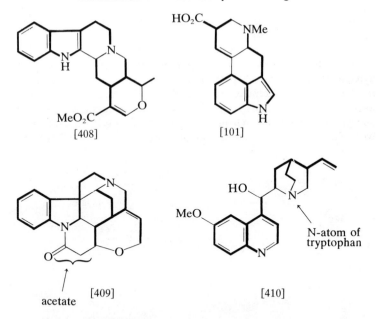

[408]   [101]

[409]   [410]

acetate

N-atom of
tryptophan

type) might be derived from an iridoid monoterpene (mentioned in
Chapter 3: see Figs 3.9 and 3.10) precursor, was the vital step that led to
the eventual elucidation of the pathways. In particular the C_{10} sub-unit of
ajmalicine-type alkaloids is very reminiscent of the iridoids loganin [114]
and secologanin [118]. Ring cleavage should produce a suitably
functionalized C_{10} precursor, which could form a Schiff base with

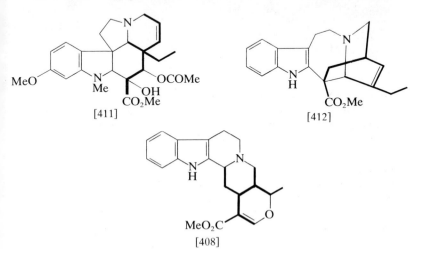

[411]   [412]

[408]

tryptamine, and thence isovincoside [119] and eventually ajmalicine [408]. This hypothetical sequence is shown in Fig. 6.13.

Indeed, recent tracer experiments carried out by Scott and Stöckigt, using cell free extracts of *Catharanthus roseus*, are in almost complete agreement with this hypothetical pathway. Scott envisages the involvement of an intermediate [413] (though this could not be isolated), as progenitor of alkaloids of the vindoline and catharanthine types [411] and [412]. A reasonable scheme is shown in Fig. 6.14. It is not difficult to write a mechanism for the formation of dehydrogeissoschizine [414], and

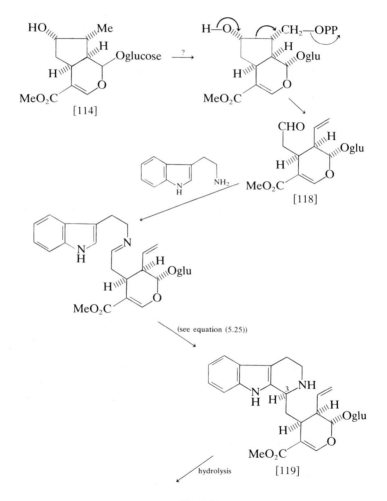

Fig. 6.13

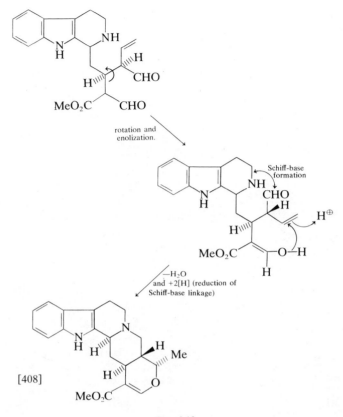

Fig. 6.13

elegant experiments carried out by Stöckigt have demonstrated conclusively that this is converted into [408]. The pathway from [414] to
hypothetical intermediate [413] is not so straightforward, and although
various schemes have been proposed these are largely speculative and we
shall not consider them further. The interested reader is urged to consult
the work by Scott and Stöckigt for a discussion of tracer experiments,
biomimetic syntheses, and of hypothetical pathways.

However, if we accept the intermediacy of an intermediate like [413], it
is interesting to consider how we can then rationalize formation of the
other types of alkaloid. Two Diels–Alder type processes, and some
functional group introduction in the case of vindoline [411], provide
rather satisfactory mechanisms for the production of vindoline and
catharanthine [412] (Fig. 6.15).

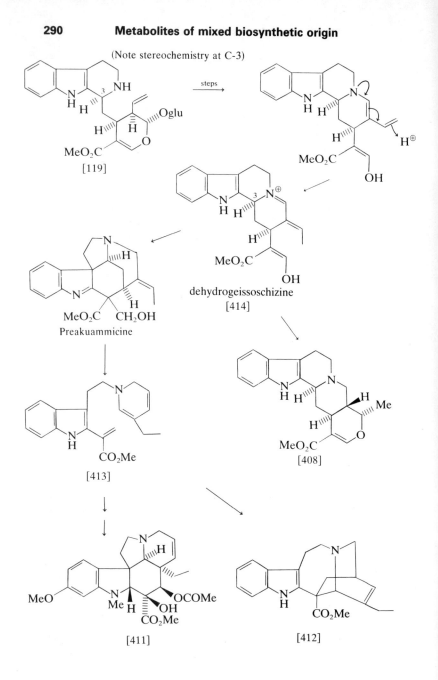

(Note stereochemistry at C-3)

steps

MeO₂C
[119]

H⊕

dehydrogeissoschizine
[414]

MeO₂C   CH₂OH
Preakuammicine

[413]

[408]

[411]

[412]

Fig. 6.14

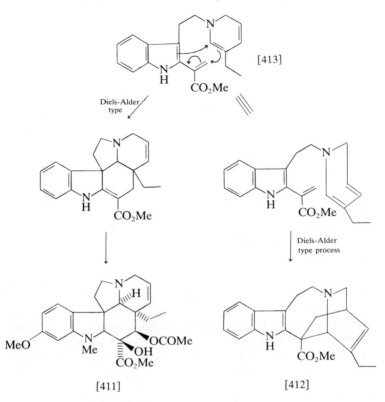

Fig. 6.15

It is worth noting that until recently (1977), vincoside (3β-epimer of [119] was believed to be the precursor of the indole alkaloids just discussed. This necessitated postulation of an epimerization step, since geissoschizine has 3-α-stereochemistry. The recent experiments of Scott and Stöckigt using cell free systems from *Catharanthus roseus* have, however, conclusively demonstrated the involvement of isovincoside (3-α epimer of [119]), otherwise known as strictosidine. Indeed the enzyme strictosidine synthase has been isolated from *Catharanthus roseus* (and several other cell cultures), and shown to have an optimum pH of 5–7.5. It has even been bound (immobilized) to activated Sepharose, and can then produce gramme quantities of strictosidine from secologanin and tryptamine!

In passing we might note that the so-called Iboga alkaloids, produced by a number of different African plants, possess the catharanthine skeleton (less the olefinic double bond). African natives chew the roots of these shrubs in order to alleviate feelings of hunger or fatigue, just as South

American Indians chew leaves of the coca plant, whose active principle is cocaine.

Finally, two, more complex, vinca alkaloids should be mentioned: vincristine [415] (R = —CHO) and vinblastine [415] (R = —Me). These

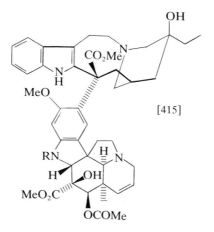

[415]

are very potent antitumour agents, and find extensive use in cancer chemotherapy. The natives of Madagascar were using extracts of *C. roseus* as a treatment for diabetes before the Eli Lilly company showed that the compounds were highly cytotoxic!

Strychnine, reserpine, and camptothecin

Precise details of the biosynthetic pathways to these metabolites are not yet available, but reasonable schemes can be proposed based upon our knowledge of the routes to the vinca alkaloids.

Strychnine [409] and the closely related brucine (dimethoxy analogue) occur in the seeds of species of the genus *Strychnos*. It is unlikely that their toxic properties are useful to these plants, but their intensely bitter taste (one part of strychnine imparts a bitter taste to 500 000 parts of water!) is of much more likely utility. Tryptophan and geraniol are both incorporated into strychnine, as expected, and the additional C_2 unit is derived from acetate. The sequence shown in Fig. 6.16 seems plausible.

The alkaloid reserpine [416], a sedative and anti-hypertensive agent, probably arises from a vinca-type intermediate as well (eqn 6.9).

The biosynthesis of camptothecin [417] has been studied recently by Hutchinson and coworkers. ([14C]-, [13C]-, and tritiated precursors were utilized, and the intermediacy of the lactam [418] was established (eqn 6.10]. It is interesting to note that, as in the case of the vinca alkaloids, only one of the two possible C-3 epimers isovincoside [119] is employed. In addition biomimetic sytheses of camptothecin have been accomplished

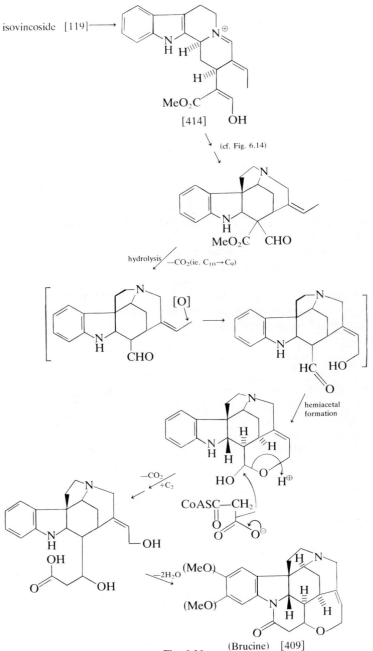

isovincoside [119]⟶

[414]

hydrolysis / —CO₂(ie. C₁₀→C₉)

(cf. Fig. 6.14)

[O]

hemiacetal formation

—CO₂ / +C₂

CoASC—CH₂

—2H₂O

(Brucine) [409]

Fig. 6.16

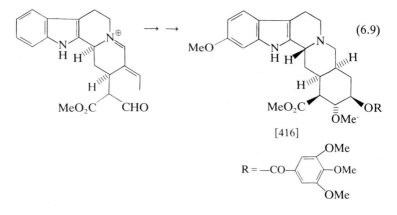

$$(6.9)$$

[416]

with greater facility when the intermediates possess 'natural' stereochemistry at C-3.

At one time it was hoped that camptothecin would prove to be a useful anti-tumour agent, but experiments with animals have been rather disappointing, and it is unlikely that it will fulfill its early promise. Attempts are being made, however, to establish which of its structural features are important for biological activity, and a number of bioactive analogues have been synthesized.

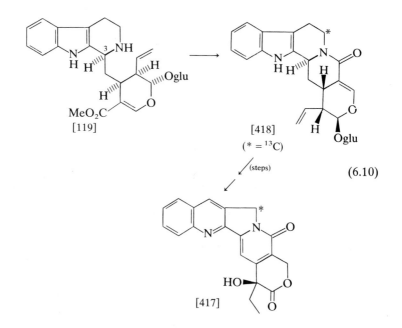

$$(6.10)$$

Quinine

Quinine [410] is of course well-known as an anti-malarial drug, and has been used for this purpose since Elizabethan times, though it has now been large superseded by safer, synthetic drugs. Its relationship to the indole alkaloids is not immediately obvious, and the tryptamine sub-unit

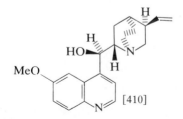

[410]

has obviously been fragmented in some way. Radio-labelled tryptophan and tryptamine have been incorporated, and the label appears as shown in Fig. 6.17. Also isovincoside is a precursor as is corynantheal [419] and although the details of the later stages of the biosynthetic pathway remain unknown, the scheme shown in Fig. 6.17 is plausible.

Ergot alkaloids

Through the ages the ergot alkaloids, which are metabolites of the fungus *Claviceps purpurea*, have achieved a certain notoriety, and for good reason. A recurrent 'pestilence' of the Middle Ages in Europe was caused by consumption of rye (usually in the form of bread) which had been infected with *Claviceps purpurea*. The symptoms of poisoning seem to have differed from one part of Europe to the other. In France, and west of the Rhine, a generalized gangrene, brought about by the vasoconstrictive properties of some of the alkaloids, was common; while in Russia convulsions and other deep-seated mental disturbances were more prevalent, and there are tales of peasants leaping *en masse* from upper storey windows, etc. This affliction was usually called St. Anthony's fire, due to the burning sensations which accompanied the convulsive type; and 'holy fire' was implicated in the blackening of limbs which was characteristic of the gangrenous type.

Of the alkaloids produced by the fungus, lysergic acid [101] and agroclavine [420] are typical, and their biosynthesis has been extensively studied, especially of late. It is well established that mevalonate and tryptophan serve as precursors of these alkaloids, and a cell-free system from *Claviceps purpurea* produces labelled chanoclavine-I [421] and agroclavine from radio-labelled tryptophan, isopentenyl pyrophosphate (IPP) [96], or from methionine. A double *cis-trans* isomerization is known to occur (note * in Fig. 6.18) and ^{13}C-enrichments were noted at the marked carbons when [*E*-methyl ^{13}C]-4-dimethyallyl-tryptophan was fed,

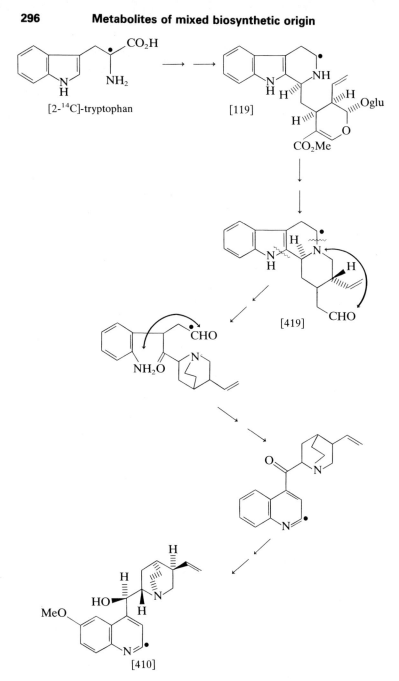

Fig. 6.17

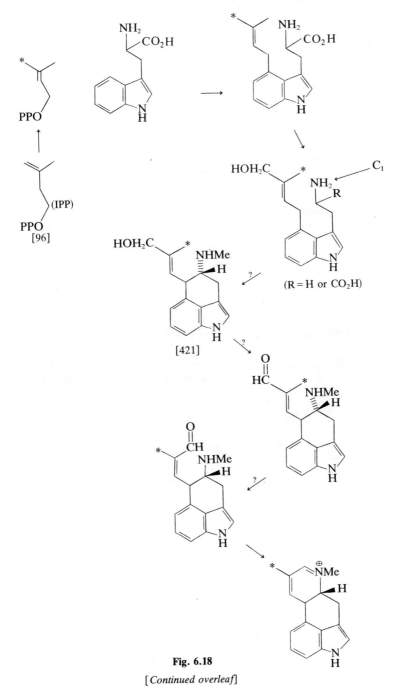

Fig. 6.18
[*Continued overleaf*]

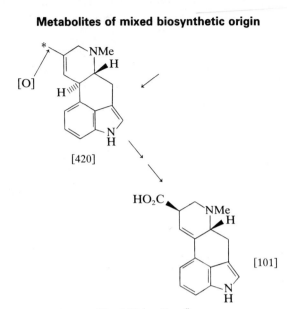

Fig. 6.18 (*continued*)

and [420] and [421] isolated. The postulated biosynthetic pathway is shown in Fig. 6.18.

In addition to the potent hallucinogen LSD, the diethylamide of [101], two other ergot alkaloids have pharmacological properties that are of

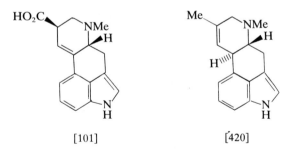

clinical utility. These are ergometrine [422a] and ergotamine [422b], which are used as uterine contracting agents following childbirth: both are amides of lysergic acid, and have been in use since the sixteenth century, albeit as crude extracts of *Claviceps purpurea*.

Finally, the most important magical preparation of the Aztecs, ololiuqui, obtained from seeds of the plant *Rivea corymbosa*, also contains derivatives of lysergic acid.

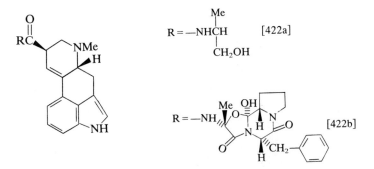

$$R = \text{—NHCH} \qquad [422a]$$

with Me and CH$_2$OH

$$R = \text{—NH} \qquad [422b]$$

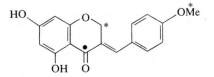

Problems

6.1. Tracer from [1-^{14}C]-phenylalanine (•) and [^{14}C]-methionine (*) was located in eucomin as shown when they were used as precursors. Account for this result.

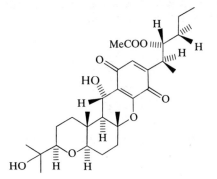

[Dewick, P. M. (1973). *Chem. Commun.*, 438]

6.2. Cochlioquinone A has the structure shown. Suggest a possible biogenesis for this metabolite.

[Beretta, M. G., Canonica, L., Colombo, L., Gennari, C., Ranzi, B. M., and Scolastico, C. (1980). *Perkin Trans. I*, 2686]

6.3. Suggest a biogenesis for the fungal metabolite α-cyclopiazonic acid. How would you establish the correctness of your proposal?

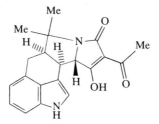

[Ferreira, N. P., Kirby, G. W., Steyn, P. S., Varley, M. J., and Vleggar, R. (1975). *Chem. Commun.*, 465]

6.4. The ginger plant produces (S)-(+)-6-gingerol, and labelling studies have indicated the following probable pattern.

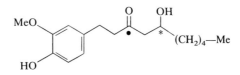

[1-¹⁴C]-phenylalanine (●)
[1-¹⁴C]-hexanoate (*)
Suggest a plausible biosynthetic pathway.
[Denniff, P. and Whiting, D. A. (1976). *Chem. Commun.*, 711]

6.5. The mycotoxin Austalide D is produced by the fungus *Aspergillus ustus*. Label from [2-¹³C]-MVA is incorporated as shown. Suggest how the rest of the molecule may arise, and propose a complete biogenetic pathway.

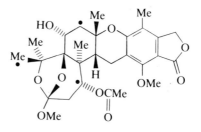

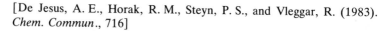

[De Jesus, A. E., Horak, R. M., Steyn, P. S., and Vleggar, R. (1983). *Chem. Commun.*, 716]

6.6. The naphthoquinone α-dunnione is present in cell cultures of the plant *Streptocarpus dunnii*. Propose a biogenetic pathway, and suggest how you might establish the validity of your proposal.

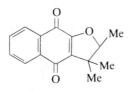

[Inouye, H., Inoue, K., Nayeshiro, H., and Ueda, S. (1982). *Chem. Commun.*, 993]

7. Secondary metabolism and ecology

In Chapter 1, the possible significance of secondary metabolism was discussed briefly. Various ideas have been considered over the years, and secondary metabolites have been thought of as products of 'overflow' metabolism (substrate levels in excess of that required for primary metabolism), or perhaps as products of minor side-routes (shunt metabolites). They may also represent detoxification products. The levels of secondary metabolites are certainly related to the stage of development of an organism, to the nutritional status, and are affected by various kinds of environmental stress. Whatever the original reason for emergence of a secondary metabolic pathway, once a useful biological function has been established, the pathways will almost certainly be retained and probably emphasized.

In the wild, every organism engages in an unceasing battle for survival as it seeks food, attempts to find a mate, and tries to avoid the attentions of its predators. Competition between different species for the same habitat or food source is common, and ultimately survival depends upon the efficiency with which an organism can compete, in an often hostile environment. However, at any one time, an unstable equilibrium exists between all species which occupy a particular ecological niche, and all are subject to sudden vicissitudes resulting from physical changes in their shared environment. In the event of a change in circumstances, be it environmental, e.g. climatic or geological, or an upset in the balance of predator to prey, any species which becomes subject to adverse pressures may adapt to the altered conditions, emigrate, or become extinct.

It may be noted that of these alternatives, terrestrial plants can only attempt to adapt (or else become extinct), since, unlike their predators, the herbivorous insects and animals, they cannot emigrate (except via seed dispersal). It is likely that this is one reason for the immense variety of secondary metabolic pathways which plants utilize. As plants moved from an aquatic to a terrestrial environment, they came into competition with an increasing number of mobile predators, initially insects, but subsequently vertebrates as well. It appears likely that plants evolved and began to utilize a number of novel metabolic pathways which produced compounds noxious to insects, and other herbivores evolved, and developed means of detoxifying or utilizing the secondary metabolites, and the plants responded with new deterrents, the complexities of secondary metabolism as we now know them, began to emerge.

Perhaps a few comments about the evolutionary process are in order. Organisms envolve through changes (mutations) in their genetic make-up, and since one gene generally specifies one protein or enzyme, concomitant changes in enzyme activity result. These genetic changes may be positive, rendering the organism more 'fit' for survival; neutral, of no apparent benefit or harm; or negative, in which case if the species is under considerable pressure, it probably takes one step closer to extinction, or disappears in the one step! Genetic change which results in an altered primary metabolic pathway is usually detrimental (though major evolutionary innovation occurs by adoption of the rare, advantageous changes). In the final analysis, a measure of the evolutionary success of an organism is provided by the number of its genes which are present in its offspring.

It is quite clear that many secondary metabolites of plant origin are used as agents of deterrence, but since evolution is a continuous process, as plants become more competitive through utilization of these compounds, so herbivores and microorganisms become tolerant of their effects, and develop ways of using them to their own ends. In addition many natural products appear to be harmful to the plant that produces them. Thus terpenes are often stored in special surface glands or in other specialized cellular compartments, many phenols occur as sugar conjugates, and other compounds, like the cyanogenic glycosides which release hydrogen cyanide on hydrolysis, are separated from the intracellular enzymes which could activate them. It seems almost as if these metabolites are an evolutionary embarrassment to the plants, but have been tolerated and accommodated because they serve a useful function.

Inherent in these assumptions is the undoubtedly controversial postulate that diversity of flora and fauna is a direct result of chemical interaction, and coadaption of plants, herbivores, and microorganisms. The relatively new science of 'ecological chemistry' seeks to identify these interactions, and to show how they affect the natural patterns and life-styles (i.e. the ecology) of the participants. It is not easy to categorize the interactions, since many involve several different species, but a few main types are apparent, and these will be discussed in turn.

Plant–herbivore interactions

In any habitat a plant interacts with plants of other species, (competitors), with herbivores (predators in the main), and with invasive microorganisms. Only the chemical interrelationships between plants and herbivores will be considered in this section, though survival may be threatened as a result of any of these associations, and also by environmental pressure, above and below ground.

Plants and their insect predators are evolving simultaneously, and usually antagonistically. The former produce new chemical deterrents or attractants (for many plants require insects for assistance with pollination): and the latter evolve new sensory capacities which allow discrimination between plant species, or produce new digestive enzymes which provide for detoxification and even utilization of these natural products.

Deterrence and attraction

The diverse associations between Danaid butterflies and their host plants exemplifies the complexity engendered by the co-evolution of insects and plants. This family (Danaidaea) includes the well-known Monarch butterfly, which has a predilection for species of plants that contain cardenolides (cardiac glycosides) such as calotropin [423]. Eggs are

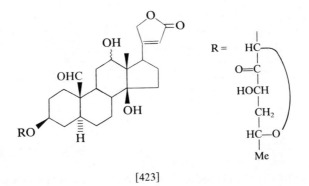

[423]

deposited on the plant, and through feeding cardenolides are assimilated during the larval stages, and incorporated into the tissues of the adult butterfly. These mature butterflies are unpalatable to most predators, though deterrence is not due to the toxicity of the cardenolides, but rather to their intense bitterness and emetic properties. Birds rarely consume more than a small portion of the insect, before regurgitating it, and discarding the remainder of the butterfly; many butterflies are thus found to have beak marks on their wings, but are otherwise sound. It is probable that cardenolides originally deterred all herbivores, but Danaids have evolved processes of detoxification and storage, and now actively discriminate in favour of plants that contain these cardiac glycosides. The secondary metabolism of the plant thus enhances the survival prospects of the insect. In addition, since these plants are avoided by the larger herbivores, the larvae also gain the advantages of cover and safety from accidental ingestion.

Danaid butterflies have accrued additional advantages through evolution: in common with other insects which can assimilate toxins, they are brightly and characteristically coloured. This aposematic (or warning) colouration is a signal to birds and other predators that the insect is distasteful; and laboratory simulations have shown that birds rarely attempt to eat aposematic insects after initial experience of their unpalatability.

Aposematism abounds, and the common ladybird provides a familiar example. The characteristic colouration (red with black spots) is a 'warning' that few would-be predators ignore. The insect synthesizes a number of toxic compounds, including coccinelline [424], and this is probably derived from a polyketide precursor (eqn 7.1).

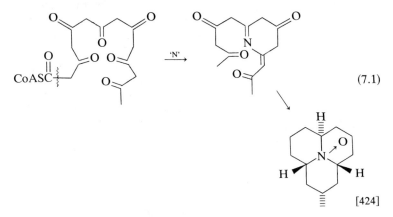

(7.1)

[424]

Before leaving the Danaids, we should note the secondary metabolites which are utilized by the male butterflies. The adult males feed upon plants (e.g. ragwort) which synthesize pyrrolizidine alkaloids, such as sennecionine [280] and can metabolize these toxic and tumourigenic alkaloids to produce the dihydropyrrolizidine ketone [284]. This ketone is

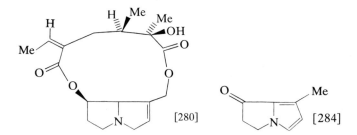

[280]   [284]

employed as a flight arrestant (and aphrodisiac?) during the prelude to mating. The male deposits a 'dust' comprising ketone [284] and a viscous component, e.g. [425] of terpenoid origin, onto the antennae of the female butterfly, and mating usually ensues.

HO〜〜〜〜CO$_2$H   [425]

The interesting suggestion has been made, that originally Danaids became dependent upon pyrrolizidine alkaloids, when the larvae of all species fed upon primitive plants which contained both cardenolides and pyrrolizidine alkaloids. These ancestral plants then evolved, in an attempt to break the pattern of insect predation, to yield at least two new plant lines, one producing cardenolides alone, and the other pyrrolizidine alkaloids. In response, the Danaids diversified such that males visited and fed upon plants containing the alkaloids, while eggs were deposited, as before, on plants containing cardenolides.

Throughout this dicussion it has been apparent that the insects derive most of the benefits, while the plants play the role of subservient host. However, plants which contain cardenolides or pyrrolizidine alkaloids are generally avoided by other herbivores, and perhaps we are witnessing a period of Danaid dominance, to be broken eventually by a further diversification of the host plants. Other plants produce deterrent metabolites which are very successful, at least against herbivores. Few insect species have yet been shown to feed upon plants containing nicotine [265]; and the pyrethrins [109] and thujaplicins [426] (ex. certain heart-woods) are also broad spectrum insect repellents. The repellent

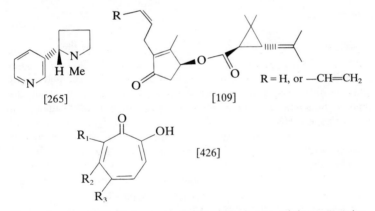

[265]   [109]   R = H, or —CH=CH₂

[426]

effects of many alkaloids is now well established, e.g. quinine, conessine, gramine, morphine, brucine, scopolamine, sparteine, strychnine, berberine, and atropine. In addition, the specific attraction of certain insects is advantageous to plants, as for example pollination of clovers by the honeybee, which is attracted to the plants by the odour of the triene [427]. In other instances the odour of a plant predator (e.g. an aphid or bark beetle) is attractive to the predators of those insects, and the plant

[427]

again benefits. Many other examples of these phenomena are known, and the interested reader is referred to the recommended texts at the end of this chapter, for further information. However, we should also consider the ways in which plants interact with the larger herbivores.

Flavour is the basis of most plant–animal interactions, though the presence of thorns and stings is also significant. Bitterness is a sensation that all mammals, and probably birds and reptiles too, can appreciate, so that plants which contain cardenolides, alkaloids, or tannins, are avoided by most vertebrate herbivores. It is worth noting in passing that primitive plants contain tannins, while more modern species (in evolutionary terms) contain alkaloids, etc., instead. Reptiles can detect tannins (and are repelled) at the same levels of concentration as mammals; but detection of alkaloids is much less efficient (10^3-fold less), and the appearance of alkaloid-containing plants may have contributed to the demise of the large, prehistoric reptiles. The volatile monoterpenes may act as insect attractants, but also serve as grazing deterrents: geraniol [90] or 2-Z-citral [149] (neral), although pleasantly odiferous to Man, may not be so

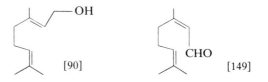

[90]   [149]

attractive when emanating from a potential food source, and are in addition bitter. The presence of caffeine or tannins in many plants also renders them less palatable, though we tolerate them in small amounts in tea, coffee, etc. However, as Haslam has noted recently, the situation with regard to tannins is probably more complex. Higher plants often have tannins with macrocyclic structures that will not easily fit into proteinaceous taste receptors. The precursors, e.g. penta-glucosylated gallic acid (see [399]), are smaller and much more suitable for interaction with a receptor. The evolution of the more complex tannins seems thus to have a negative correlation with astringency.

There are of course examples of animals which have developed special digestive or detoxification enzymes to cope with toxic plant metabolites, and other means by which to overcome the physical barriers like thorns and stings. Thus koala bears consume large quantities of Eucalyptus leaves, which are rich in terpenes and phenols; a species of mouse (*Mus musculus*) feeds avidly upon milkweed and Monarch butterflies, both of which contain cardenolides; and the white-tailed deer consumes the alkaloid-rich mountain laurel and rhododendron.

The highly developed olfactory and gustatory senses of vertebrates allow them to discriminate between palatable and unpalatable plants, and this discriminatory process is another example of an ecological interaction mediated by secondary metabolites.

Hormonal effects

A more subtle, hormonal interaction occurs between certain plants and insects. We noted in Chapter 3 the complex interplay of juvenile hormones and ecdysones in the moulting and metamorphosis of insect larvae; and it is by interference with this delicately balanced hormonal system, that some plants achieve a competitive advantage.

The balsam fir, *Abies balsamea*, synthesizes (+)-juvabione [428] (so-called 'paper factor', *vide infra*), which, at least in some insects, disrupts the normal process of larval maturation, and metamorphosis does not occur. Juvabione appears to act as a juvenile hormone [132a] mimic, and larvae feeding upon the balsam fir are assured a continual supply of artificial juvenile hormone. Since metamorphosis occurs when juvenile hormone production ceases, the larvae cannot metamorphose, and giant larval forms are produced, which die.

The discovery of this phenomenon provides an illuminating example of how a chance observation can lead to major advances in our knowledge of Nature. A common European bug, *Pyrrhocoris apterus*, was reared on paper towels, and failed to undergo metamorphosis. However, when raised on Whatman filter paper, normal maturation occurred. Subsequent trials with American newspapers and journals also led to aberrant development, and the source of this 'paper factor' was eventually traced to the balsam fir, which provides much of America's paper. The contemporaneous discovery of natural juvenile hormones, e.g. [132a], led

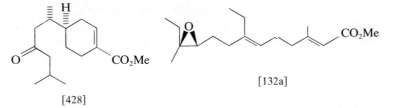

[428]   [132a]

to an understanding of this phenomenon, and also to the synthesis of several totally artificial juvenile hormone analogues, which have some potential as environmentally acceptable insecticides. Numerous plant species have been screened for compounds like juvabione, and many do possess metabolites with hormonal activity: a recent example is compound [429] from Avocado leaves. These plants presumably gain a competitive advantage since the number of adult insects in their vicinity should rapidly diminish following an initial colonization.

The other type of hormonal interaction concerns the ecdysones. Insects almost invariably rely upon plants for their supply of the steroid nucleus,

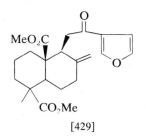

[429]

usually in the form of ingested phytosterols, which are metabolized to yield cholesterol [87] (or its equivalent), and thence into their essential structural and hormonal steroids. The ecdysones, such as α- and β-ecdysones [183] (R = H and OH), are required for correct larval development, and in particular seem to control the process of moulting. In

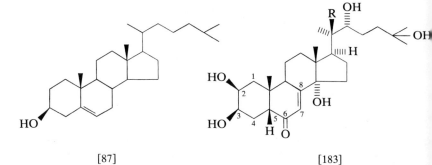

[87]   [183]

this context it is interesting to note that certain crustaceans also have a requirement for ecdysones, e.g. callinecdysone-B [430] (soft-shell crab), and β-ecdysone (crayfish): a fact which should not surprise us since insects and crustaceans are derived from the same primaeval line.

Many plant species also synthesize ecdysones, and although there is little evidence, this biosynthetic capability may provide competitive

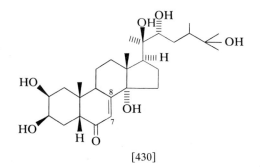

[430]

advantages for these plants through interference with the endocrine systems of insect feeders.

The subtlety of some of these ecological interactions is further enhanced by other factors, such as the association of insects with symbiotic microorganisms. Thus metamorphosis of the beetle *Xyleborus ferrugineus* will only occur if the larva receives the metabolic assistance of the symbiotic fungus *Fusarium solani*. This latter organism has the capacity to introduce a 7–8 double bond into the nucleus of dietary sterols: such functionality is an essential feature of all ecdysones (see [183] and [430]). Aposymbiotic larvae (i.e. symbiont-free) will only metamorphose if provided with dietary ergosterol [179], which already

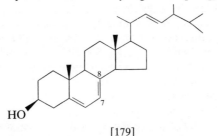

[179]

possesses this structural feature. This example, and others like it, provides a salutary reminder that before tampering with delicate ecological systems by the introduction of synthetic insecticides, etc., we should first examine their complexities. In this particular instance an anti-fungal compound might be more efficacious than an insecticide due to greater specificity of action.

Finally, some plant metabolites are known to reduce the fecundity of mammalian species: mimosine [431] causes irregular oestrus and complete infertility in laboratory rats, and cycasin [432], from the cycad nut, reduces the litter size if consumed prior to mating. The significance of such compounds in the wild is not known, but it is likely that plants do control the size and diversity of animal populations, and this possibility is often overlooked.

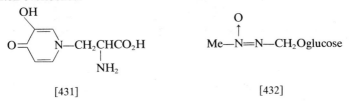

[431]   [432]

Insect–insect interactions

Insects utilize a multitude of small, volatile organic compounds as mediators of intra- and interspecies interactions. These compounds are typically monoterpenes, benzoquinones, simple phenols, fatty acid esters, and the products of their metabolism. They may be the products of *de novo* biosynthesis, but are more commonly derived by modification of dietary terpenes, phenols, and fatty acids of plant origin. Several fundamental roles for these metabolites may be distinguished: sexual attraction, aggregation factors, inducement to forage, alarm signals, and repellency. Where compounds are disseminated for reception by insects of the same species (intraspecies interactions), they are known as *pheromones*.

Sex pheromones
Many adult insects (usually females) attract a mate, often from a considerable distance, by releasing sex pheromones. These are most commonly aliphatic compounds of long chain length, and are mostly derived via metabolism of fatty acids. Some typical examples are shown in Fig. 7.1. Most sex pheromones are completely species-specific, and in this way unfruitful attempts at cross-mating are avoided. Where two different species utilize the same compound, either additional pheromones are present, providing a different, more complex signal, or the two species will mate at different times of the day or year, i.e. they are not receptive concurrently.

It is also notable that precise structural specificity is often essential. Small structural modifications produce profound changes in biological activity. Thus compound [433] is the natural pheromone of the redbanded leaf roller (insect names are often as exotic as their pheromones!); while the saturated ester [434] is a synergist, i.e. it enhances the activity of the natural pheromone; and [435], which is merely the geometric isomer of the natural attractant, inhibits sexual attraction caused by [433].

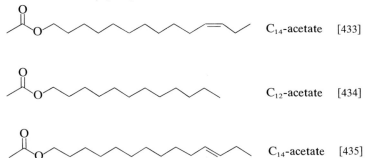

C_{14}-acetate   [433]

C_{12}-acetate   [434]

C_{14}-acetate   [435]

Aggregation factors
As noted previously insects usually have a predilection for particular plant species as a food source or as a site for egg deposition. Insects are

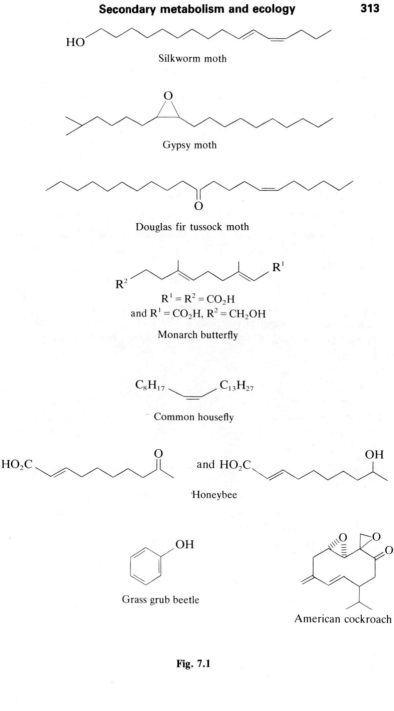

HO

Silkworm moth

O

Gypsy moth

O

Douglas fir tussock moth

R^1
R^2

$R^1 = R^2 = CO_2H$
and $R^1 = CO_2H$, $R^2 = CH_2OH$

Monarch butterfly

C_8H_{17} $C_{13}H_{27}$

Common housefly

HO_2C O   and HO_2C OH

Honeybee

OH

Grass grub beetle

O   O   O

American cockroach

Fig. 7.1

attracted to these host plants by the volatile compounds which they release: often monoterpenes, but also such compounds as anethole [231], and anisic aldehyde [232] (from the citrus and parsley families), or sinigrin [436] and allylisothiocyanate [437] which are amino acid

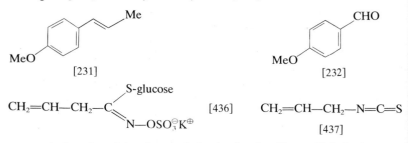

[231]   [232]

$CH_2{=}CH{-}CH_2{-}C$ S-glucose / $N{-}OSO_3^{\ominus}K^{\oplus}$ [436] $CH_2{=}CH{-}CH_2{-}N{=}C{=}S$
                                                                                      [437]

metabolites found in plants of the family Cruciferae. Sinigrin is not volatile, but is nonetheless attractive to aphids and other insect species, as it stimulates feeding.

The subsequent pattern of events varies from species to species, and we shall consider a few typical examples. The male cotton boll weevil (*Anthonomus grandis*) is attracted to the cotton plant by the odiferous mono- and sesquiterpenes, primarily α-pinene [159], limonene [438], and caryophyllene [125], which it releases. He feeds, and then presumably

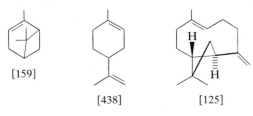

[159]   [438]   [125]

utilizes these metabolites for elaboration of his own aggregation pheromones: these are attractive to both males and females. Four biologically active compounds have been isolated and characterized, [155]–[158], and in field trials, chemically synthesized pheromones have been used successfully to attract large numbers of boll weevils for extermination. Probably the best example of a large-scale trapping

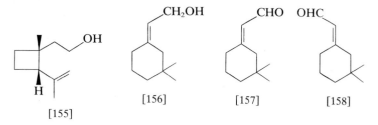

[155]   [156]   [157]   [158]

campaign was that carried out in Scandinavia in 1979/80. A synthetic pheromone mixture specific for the spruce bark beetle (*Ips typographus*) was employed, and the total catch of beetles was 2.9 billion in 1979, and 4.5 billion in 1980 (solely in Norway!).

This train of events is typical, and offers obvious ecological advantages to the insects: mating occurs on or near the host plant, and the larvae are assured a source of food. The plant, however, appears to derive little benefit, as the insects exploit its 'defence system' by utilizing its 'deterrent' natural products.

A more complex series of interactions occur between bark beetles, which are a major scourge of forests everywhere. A female of the species *Dendroctonus brevicomis* is attracted to their host tree by the resin exudate: mainly β-myrcene [439], β-pinene [440], and car-3-ene [441].

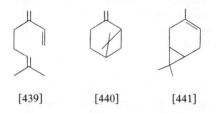

[439]   [440]   [441]

She then releases brevicomin [442] and frontalin [443] which are attractive predominantly to males and females respectively: approximately equal numbers of both sexes are thus caused to aggregate. Both aggregation factors are products of fatty acid metabolism. The closely related species. *Dendroctonus frontalis* utilizes a different strategy. Here, frontalin [443] is primarily attractive to male beetles, and the female does not release a female attractant. To avoid excessive male response, the first male arrivals ingest resin, then synthesize and release a male inhibitor,

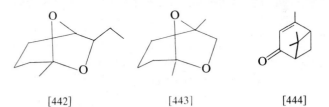

[442]   [443]   [444]

verbenone [444]. It seems likely that the male beetles have the capacity to convert α-pinene [159] (the major attractant in the tree resin) into verbenone. Another species of bark beetle, *Ips paraconfusus*, certainly does have a similar capability: a symbiotic gut bacterium, *Bacillus cereus*, metabolizes α-pinene to produce *cis* and *trans*-verbenols [445] which are female attractants for this species.

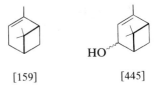

[159]   [445]

Trail pheromones
Trail pheromones are used almost exclusively by social insects, such as ants, termites, and bees. An insect which has located a new source of food, returns to the nest, marking his trail by releasing these substances. At the nest recruitment is initiated by perception of these trail pheromones, and of other recruitment pheromones which the insect now releases, or by physical contact between the successful forager and other members of the colony he wishes to recruit. Typically, for social insects, many of these interactions, physical and chemical, are exceedingly complex and well-developed. Some representative trail pheromones are shown in Fig. 7.2.

Alarm pheromones
As their name suggests, these compounds are used as warning signals, and advise other members of the species of imminent danger. The common bed-bug releases *E*-hex-2-en-1-al and *E*-oct-2-en-1-al [446] (*n* = 2 and 4); while ants use a miscellany of compounds, including citral (both geometric isomers: neral and geranial) [149], undecane, tridecane, and several cyclopentanones. The ant *Atta texana* uses the heptanone [447].

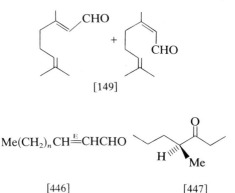

[149]

$$Me(CH_2)_n CH\overset{E}{=}CHCHO$$

[446]   [447]

Most of the compounds considered thus far are used for long-range communication between members of the same species. In contrast, the defensive secretions and sprays are agents of short-range interactions, and are usually broad spectrum deterrents.

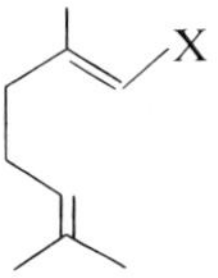

$X = -CH_2OH, -CHO, -CO_2H$
geraniol, geranial, geranic acid

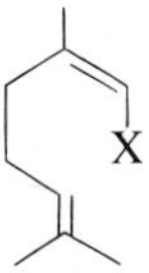

$X = -CH_2OH, -CHO, -CO_2H$
nerol, neral, nerolic acid

Honeybee: appear to control swarming

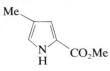

Southern subterranean termite (of dietary
origin — ex. fungus-infected wood)

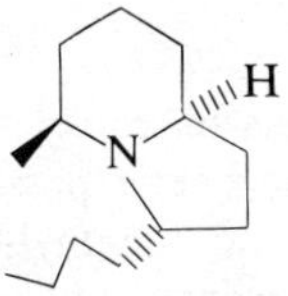

Pharaoh ant

Leaf-cutting ant
(*Atta texana*)

Fig. 7.2

Defensive compounds
Nowhere has the art of 'chemical warfare' been developed to such a
degree as in the insect kingdom. All of the common pathways of
secondary metabolism are used by one species or another (though dietary
sources are also important) for the elaboration of a veritable arsenal of
obnoxious chemicals. These may be exuded from special glands, sprayed,
or injected, but may also be present in blood and tissues to render the
insect unpalatable (cf. the Monarch butterfly).

The spider *Vonones sayi* secretes a mixture of quinones [488], (R = H or Me; acetate pathway?), which are both crystalline at ambient temperatures, but exist as a liquid in the proportions found in the spider: an example of a mixed melting point depression!

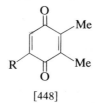

[448]

The whirligig beetle is not only superbly equipped to flee from predators — its hindlegs can reach 50–60 strokes per second; but it also has an impressive chemical defence mechanism. The compound [449],

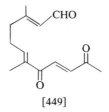

[449]

gyrinal, is a powerful antiseptic (prevents microbial attack) and is also toxic to fish and mammals. Finally, the beetle possesses a distinctive warning odour due to 3-methyl-butanal and 3-methyl-butanol.

The soldier beetle, *Chauliognathus lecontei*, utilizes the acid [450] (acetate pathway), while the ant, *Iridomyrmex*, employs iridodial [113]

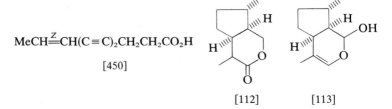

$$MeCH \overset{Z}{=\!\!=} CH(C \equiv C)_2CH_2CH_2CO_2H$$

[450]

[112]   [113]

and iridomyrmecin [112] in its defensive secretions (mevalonate pathway). A more exotic compound is employed by a Cuban termite: the nitro olefine [451].

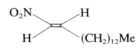

[451]

It is noteworthy that many defensive exudates are slimy or sticky, or become so on exposure to the air, and they tend to immobilize the mouthparts of predatory insects: an obvious advantage for otherwise helpless creatures like slugs and worms.

Several defensive sprays contain formic acid or quinones, and the insects that use this means of defence (primarily beetles) are often excellent marksmen. Residual traces of the deterrent compounds remain on the insect's body after spraying, and serve to reinforce the effects of the initial assault: this may be of particular utility if the insect is under mass attack, by, for example, ants.

One of the most exotic defence mechanisms is employed by the bombardier beetle. The beetle possesses internal storage glands which contain a mixture of hydroquinone and methyl-hydroquinone [452] (R = H and Me), and hydrogen peroxide, in aqueous solution. Adjacent,

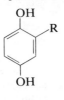

[452]

but peripheral, glands contain a mixture of enzymes including catalases and peroxidases. In response to attack, the reagents are forced into the peripheral 'reactor' glands, and a violent reaction ensues resulting in the formation of quinones and oxygen. The quinones are then released with explosive violence, as a hot (as much as $100°C$), gas-propelled spray. There is even said to be an accompanying sound like a pistol shot. Early monastic investigators reported: 'You can hear the pistol and smell the gunpowder'. All in all an alarming experience for a would-be predator.

Injected venoms are important primarily as killing agents, but may also be used in defence. Examples are formic acid. and the alkylpiperidines [453] from the fire ant, which by analogy with the Hemlock alkaloids, are probably of polyketide origin.

$$Me \overset{\displaystyle\frown}{\underset{H}{N}} (CH_2)_n Me \qquad n = 10,\ 12,\ \text{and}\ 14.$$

[453]

Many insects have also developed complex formulations which facilitate penetration of toxic or irritant components. These include wetting agents (e.g. octanoic acid), and compounds which soften the hard, insect cuticle, e.g. nonyl acetate.

Finally, insects may, like the Monarch butterfly, simply be unpalatable, and the blister beetle (family Meloidae) is exemplary. This contains considerable amounts of cantharidine [454], ('Spanish fly'), a powerful

[454]

irritant, and once thought to possess aphrodisiac properties: given orally it often produced fatal results, rather than those intended. It has been established that cantharidine is biosynthesized from MVA by male beetles: females do not have this capability. During mating cantharidine is transferred from the male to the female, and mating stimulates *de novo* synthesis of the compound by the male beetle.

The list of defensive compounds and defence strategies is almost endless, and lest the reader is tempted to believe that such 'chemical warfare' is confined solely to the insect kingdom, we should consider a few examples from higher organisms. South American frogs are a particularly good source of novel, toxic metabolites, a finding upon which the local Indians capitalize: the compounds are used as arrow poisons. Thus a Colombian frog produces batrachotoxin [194], while other South

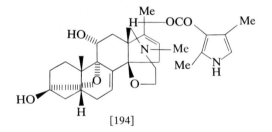

[194]

American frogs use histrionicotoxins e.g. compound [455], and salamander skin contains the deadly neurotoxin, samandarin [193]. The orange-speckled frog produces a skin secretion which contains 5-hydroxytryptamine (10% of the dry weight) [342], which induces sneezing and catarrh-like symptoms in Man.

The one unifying feature which underlies all of these interactions is the competitive advantages which accrue to any creature which can successfully defend itself, by whatever means. We see here, as elsewhere, the results of adaption through evolution, and the associated production of new biosynthetic capabilities or means by which ingested plant metabolites may be used to the advantage of the consumer. In either

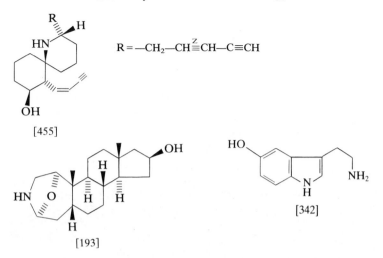

$$R = -CH_2-CH \overset{Z}{=} CH-C \equiv CH$$

[455]

[193]

[342]

instance, successful adaption to external pressures is associated with increased competitiveness.

Plant–plant interactions

Plants interact with each other no less vigorously than insects, but these interactions are generally non-specific. Secondary metabolites are released into the environment, above ground from foliage, tree resin, etc., or below ground via the roots; and these compounds reduce competition from other species by inhibiting their germination or growth. This is *allelopathy*. Often self-toxicity is evident, but the associated benefits usually outweigh the disadvantages, especially for species of short life-times. Again the whole gamut of secondary metabolism is represented, and all compounds are synthesized *de novo* within the plants concerned.

Above ground two methods of dispersal are employed: rain leaching and volatilization. In arid climates, volatile compounds, usually monoterpenes, are used almost exclusively; while in very humid environments (e.g. rain forests) phenols and other water-soluble species are utilized. As might be anticipated, in places where the climate is intermediate between these extremes, a variety of volatile and leachable allelopathic agents are involved.

Some idea of the effectiveness of allelopathy can be discerned from consideration of the ecology of the semi-arid deserts of southern California. These areas are dominated by two species of shrub: *Salvia leucophylla* and *Artemesia californica*, the so-called chaparral. During the summer the air is heavy-laden with the scent of monoterpenes, primarily

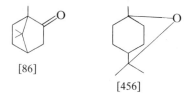

[86]

[456]

camphor [86] and 1,8-cineole [456] which emanate from these plants. The terpenes fall to earth and are absorbed onto soil particles, leading to inhibition of germination in the following spring. Those plants which do grow, suffer respiratory impairment due to air-borne terpenes. It is only in the wake of a bush fire (a common occurrence in the chaparral where the air is saturated with inflammable compounds), with concomitant destruction of soil- and air-borne monoterpenes, that other species of plants appear in any numbers. This 'fire-cycle' of the chaparral, with ascendancy, then dominance, and final destruction by fire, is a major factor in the overall ecology of these Californian deserts.

Many phenols are broad spectrum allelopathics, including *p*-hydroxybenzoic acid, vanillic acid [249], ferulic acid [250], and *p*-coumaric acid [219]: all produced via the shikimate pathway. The

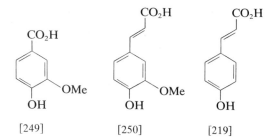

[249]   [250]   [219]

shikimate metabolite, juglone [56], is responsible for the restricted growth of plant species under and around walnut trees: it occurs *in vivo* as the corresponding hydroquinone (possibly as a glycoside), and this is rain-leached from the leaves and oxidized to juglone in the soil. Most phenols are stored as their water-soluble glycosides, and are thus more easily leached by rain.

[56]

An interesting example of how species may become tolerant of the effects of these allelopathic agents, is provided by the Eucalyptus. This was introduced into the United States, and few other native species were able to grow in its vicinity due presumably to the allelopathic effects of the terpenes and phenols that it produces. In Australia, however, where the Eucalyptus is indigenous, this dominance has been overcome by many species of plants. This is surely a consequence of the increased time-span during which Antipodean plants have been able to evolve such that they can compete with the Eucalyptus for the same habitat. Presumably, in due course, American species will do the same.

Release of allelopathic agents from plant roots is also common, and a few examples will suffice. *Nicotiana tabacum* releases nicotine [265], and fields of barley are usually relatively weed-free due to the effects of hordenine [298] and gramine [346]. Gramine also deters feeding by

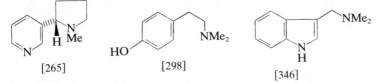

[265]   [298]   [346]

mammalian herbivores, and decreases susceptibility to aphids. In this context it is worth noting that certain tropical grasses gain immunity from insect attack (e.g. from the scourge of locusts and other infestations) because they synthesize terpenes: for example lomongrass contains high concentrations of citral. Non-tropical grasses, in contrast, are not subject to insect predation to any extent, and have merely evolved means of increasing their competitiveness with other species of plants. Two other examples of allelopathic agents released from roots are phlorizin [387] from the apple tree, and scopoletin [222] from the oak.

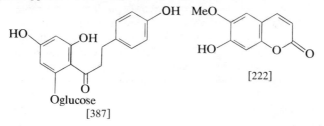

[222]

Oglucose
[387]

Finally, one type of specific intraspecies interaction should be mentioned. This is the utilization of so-called gamones by marine plants and fungi. These are essentially marine sex pheromones, usually released by the female plant or fungus, and attractive to male gametes. Typical examples are sirenin [457] and antheridiol [458] both from water moulds, and compounds [459]–[461] from Hawaiian seaweeds.

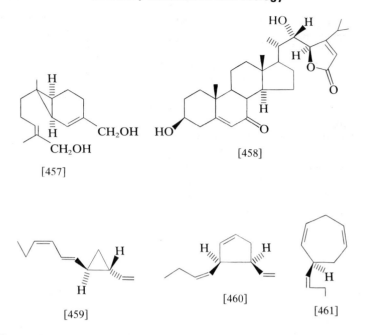

[457]

[458]

[459]

[460]

[461]

Plant–microorganism interactions

Many plants respond to microbial invasion by producing compounds which inhibit the growth of the pathogen: these substances are known collectively as *phytoallexins*. They are generally absent from the healthy plant, and actual cellular damage is a common prerequisite for initiation of their biosynthesis. Other kinds of environmental stress (UV light, heavy metal ions and toxic chemicals) also induce the formation of phytoallexins. Most classes of natural products are represented, illustrating once again that any of the basic metabolic pathways may be adapted to produce structurally diverse compounds with the same biological effect. It is a notoriously difficult area of research, since the plants usually produce small quantities of phytoallexin, and only for a limited period. Nonetheless, with the advent of modern spectroscopic techniques, recent progress has been rapid, and several hundred phytoallexins have been characterized. Some of these are shown in Fig. 7.3.

Other interactions mediated by secondary metabolites

Our discussion thus far has centred around the interactions of insects and plants, but other organisms also attempt to coexist, or actively compete with one another, and many of these interactions are also mediated by secondary metabolites. Much work remains to be done, and we can only

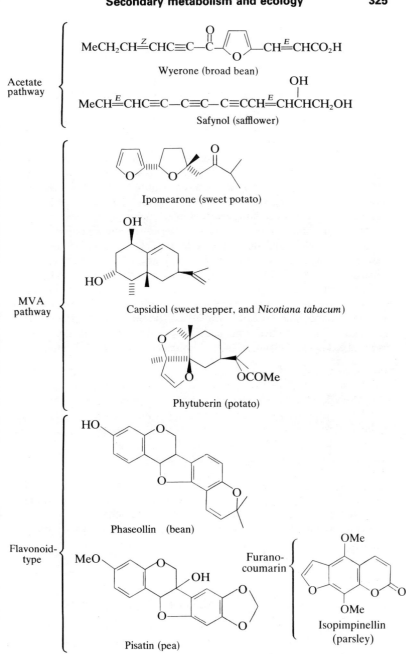

Acetate pathway

Wyerone (broad bean)

Safynol (safflower)

MVA pathway

Ipomearone (sweet potato)

Capsidiol (sweet pepper, and *Nicotiana tabacum*)

Phytuberin (potato)

Flavonoid-type

Phaseollin (bean)

Pisatin (pea)

Furano-coumarin

Isopimpinellin (parsley)

Fig. 7.3

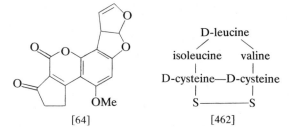

[64]   [462]

speculate as to whether production of aflatoxin [64], increases the competitiveness of *Aspergillus parasiticus*; or whether malformin A₁ [462] from *Aspergillus niger*, which has a multitude of effects including induction of malformations in higher plants, has any ecological significance. It is known, however, that many mammals use pheromones in much the same way as insects. Species-specific sex-attractants, territorial markers, and dominance odours, have been discovered but have not been studied extensively. Female pigs become sexually excited when they smell the odour of the boar salivary metabolite [463]; the striped hyena uses the α-diketone [464] as a territorial marker: the offensive odour of the skunk is due to but-2-ene-l-thiol and 4-methylbutane thiol (66% of the volatiles) and other thiols (offensive to its enemies and attractive to its mate?); while the Marten family exude [465]. The tail scent glands of the red fox contain the degraded terpenoid

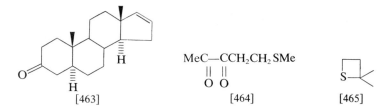

[463]   [464]   [465]

[466]; the scent gland of the Canadian beaver contains at least fourteen nitrogen-containing compounds, notably (−)-castoramin, [467], and the urine of the male house mouse contains the pheromone [468] which is related in structure to the bark beetle pheromone brevicomin [442]. It has

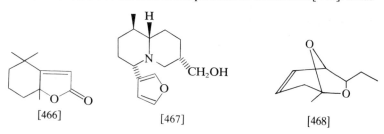

[466]   [467]   [468]

also been established that rhesus monkeys (and other primates) utilize a series of short chain-length fatty acids as sex pheromones; and there is some suggestion that similar 'pheromones' may be exuded by human beings. Males certainly secrete [463].

It is interesting to speculate about the possible significance of such compounds. It is unlikely that they have any effect in the presence of deodorants and perfumes, but surely in some earlier epoch, when Man was little more than an advanced ape himself, the odour of sweat released in moments of sexual excitement, fear, or anger, may have served as a signal of sexual attraction, or alarm, or even as a means of deterrence. Man has no natural enemies or competitors, except for himself, and it is perhaps not surprising that he produces few secondary metabolites, since these appear to mediate interactions between less dominant organisms that must increase their competitiveness. Higher organisms have attained a form of superiority through development of a nervous system and brain, but in this process of evolutionary change, they have sacrificed the ability to elaborate a range of secondary metabolites. Thus it appears that the diverse behavioural patterns of higher organisms replace the diverse array of secondary metabolites utilized by lower organisms; and surely both systems are still evolving. It is a sobering thought that an environmental cataclysm might only spare the lower organisms which retain the capacity for chemical ingenuity. The meek might indeed inherit the earth!

Finally, although all of the 'chemical communication' just described occurs on the land, much is also known about interaction between marine organisms. Two examples must suffice. The pavoninins, for example, pavoninin I [469], are produced by the sole and act as extremely effective

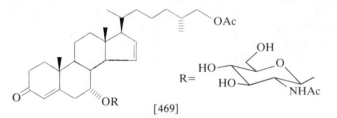

shark repellents. The fish can discharge as much as 70 mg of a mixture of pavoninins at the last moment of an attack, thus avoiding too much dilution by the seawater. Marine sponges have proved to be a rich source of secondary metabolites, and the dibromotryptamine [470] from *Smeno-*

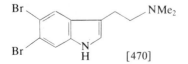

spongia aurea, possesses marked antimicrobial activity, which may be one means of defence for these sedentary, filter-feeding creatures.

This book has been primarily concerned with basic secondary metabolism, and it is only in this last chapter that we have considered possible reasons for the evolution of these processes. This bias in subject content reflects the present state of our knowledge. Time and again we have noted how the same biological effect (e.g. deterrence, sexual attraction, inhibition of growth, etc.) is achieved through utilization of natural products of diverse biogenetic origins. Surely the only explanation for such diversity is that any mutation (i.e. change in the genetic make-up) which breaks the generally harmful predator–prey relationship, or otherwise increases the competitiveness of an organism, will be favoured. In this way an initially simple pattern of secondary metabolism has become more complex through evolutionary time. The products of such evolution are the new pathways of biosynthesis, and secondary metabolites which mediate novel inter-and intraspecies interactions. The study of ecological chemistry is still in its infancy, and there is almost limitless scope for those who would partake in research in this area.

We live in an age when economics and the social conscience dictate that to be worthy of study, a problem should be relevant to our present situation, and of potential utility to Mankind: it is no longer sufficient that it is interesting. The still unknown ecological interactions which are mediated by natural products are manifold, and provide an excellent *raison d'être* for research in this area, as well as for the study of secondary metabolism itself. We may never completely understand the forces at work in the world, but it is exciting to attempt the impossible, and an increased awareness of the chemical complexities of our environment can do nothing but good. Man is an 'unstable dominant' on this Earth, and he often interacts capriciously with it. Perhaps when we have learnt more about our world, our assaults upon the planet will become more rare, and will be felt less keenly by its inhabitants, including ourselves.

Solutions to problems

2.1. Three possible modes of assembly are shown below:

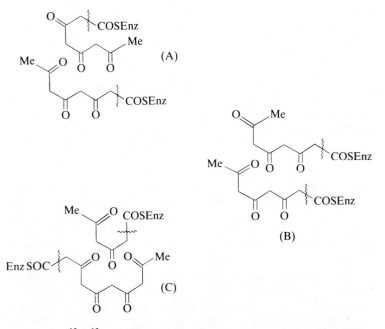

The observed $^{13}C-^{13}C$ coupling data is most consistent with a biogenesis via folding pattern C, since both patterns A and B would require coupling between C-7 and C-11, and not between C-6 and C-7.

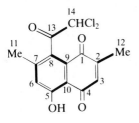

2.2. The biogenesis of metabolite A (5-chloro-3,4-dihydro-8-hydroxy-6-methoxy-3-methylisocoumarin) is straightforward:

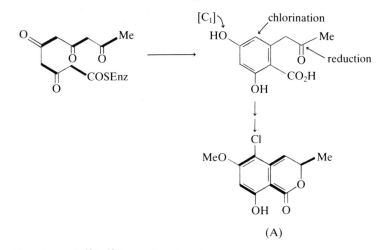

(A)

The observed ^{13}C–^{13}C coupling data fits this perfectly. Holker and Young envisaged the rearrangement shown below for the biogenesis of metabolite B.

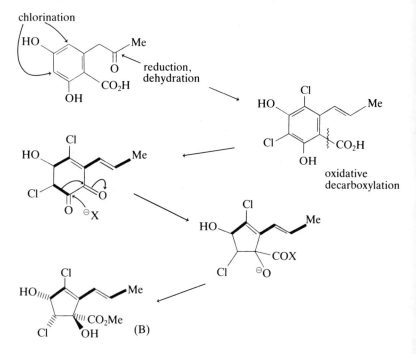

(B)

This pathway is again consistent with the labelling data, including the existence of three residual pairs of coupled carbon atoms.

More recently, Henderson and Hill have shown that the dihydroisocoumarins I, II, III, and IV are efficient precursors of B, thus establishing that chlorination occurs early in the biosynthetic pathway.

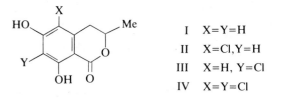

I   X=Y=H

II   X=Cl, Y=H

III   X=H, Y=Cl

IV   X=Y=Cl

[Henderson, G. B. and Hill, R. A. (1982). *Perkin Trans. I*, 3037]

2.3. The data obtained by Staunton and Evans are consistent with the coupling pattern shown below:

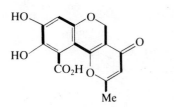

Citromycetin

The metabolite fusarubin co-occurs with citromycetin, and Hutchinson and coworkers have suggested the pathways shown below for these two related metabolites.

[Kurobane, I., Hutchinson, C. R., and Vining, L. C. (1981). *Tetrahedron Lett.*, 493]

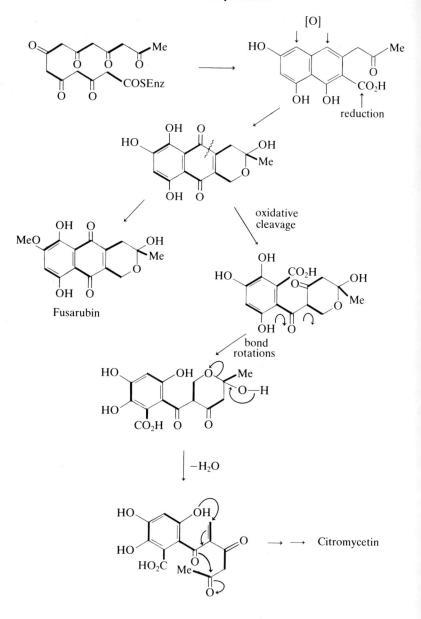

Fusarubin

reduction

oxidative
cleavage

bond
rotations

$-H_2O$

Citromycetin

2.4. In the main, benzophenones of fungal origin (like ravenelin) are produced via the polyketide pathway, while those found in plants are derived from shikimate and polyketide moieties. The data obtained by Birch are consistent with the pathway shown below:

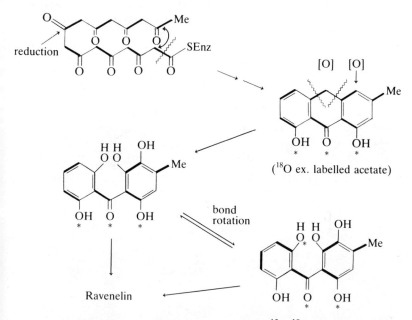

(^{18}O ex. labelled acetate)

Ravenelin

bond rotation

More recently, Vederas and coworkers have used [1-^{13}C, ^{18}O$_2$]-acetate to probe the biogenetic pathway, and since the ^{18}O-isotope caused shifts in the ^{13}C signals for C-1, C-8, C-9, and C-10a, this is also consistent with the pathways shown above (see asterisks).

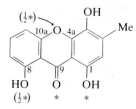

[Hill, J. G., Nakashima, T. J. and Vederas, J. C. (1982), *J. Am. Chem. Soc.*, **104**, 1745]

2.5. On the basis of the labelling studies with $[1,2\text{-}^{13}C]$-acetate, Holker and Simpson proposed a biogenesis from a polyketide, with subsequent rearrangement and decarboxylation. The two carbons marked with an asterisk were originally part of the same $^{13}C\text{-}^{13}C$ bond, and in the final product (aspyrone) they bear a 1,3-relationship to one another. In consequence, a small coupling might be anticipated, and was indeed observed. (6.2 Hz), thus supporting the mechanism proposed.

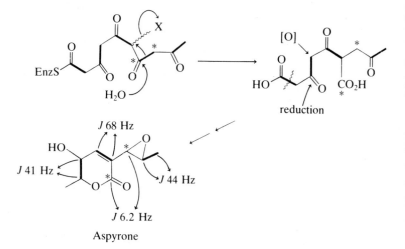

Aspyrone

More recent work by Staunton and coworkers, concerning the biosynthesis of the co-occurring metabolites asperlactone and mellein, has confirmed that a polyketide intermediate is converted into the two unsaturated lactones (asperlactone and aspyrone) without the intermediacy of an aromatic species. They have proposed a scheme which includes both lactones and mellein, and which is consistent with the observed 1,3-coupling. This is shown below.

[Staunton, J., Brereton, R. G., and Garson, M. J. (1984). *Perkin Trans. 1*, 1027]

[Ahmed, S. A., Simpson, T. J., Staunton, J., Sutkowski, A. C., Trimble, L. A., and Vederas, J. C., (1985). *Chem Commun*, 1685]

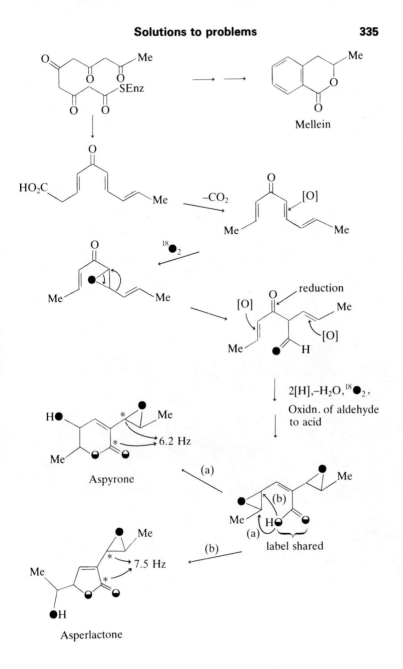

2.6. A reasonable polyketide precursor is shown below, together with the anticipated labelling pattern. Isotopically induced shifts for those carbons marked with an arrow were in fact observed when the ^{13}C-n.m.r. spectrum was taken of a sample of monocerin produced in a feeding experiment using [1-^{13}C, ^{2}H$_3$]-acetate. Complimentary results were obtained with [1-^{13}C, ^{18}O$_2$]-acetate. Several possible intermediates have been proposed on the basis of the labelling results.

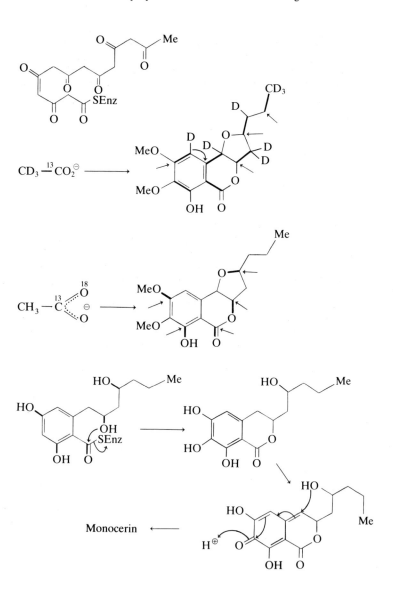

2.7. The $^{13}C-^{13}C$ coupling data is compatible with the intermediacy of the polyketide shown below:

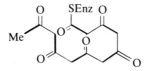

As to the timing of subsequent reactions, incorporation of $CD_3CO_2^\ominus$ provided a sample of the metabolite enriched in deuterium at C-10, C-5, C-2, and C-3. There was no enrichment at C-4. Involvement of a naphthalene derivative seems likely, and this is shown below. Several features of the labelling pattern are of note.

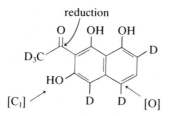

Firstly, there is little likelihood of a deuterium atom reaching C-3, unless an NIH shift reaction has taken place when hydroxylation occurs at C-4. In addition, the level of deuterium enrichment at C-10 was lower than anticipated, implying there was retention of the C-9 carbonyl until near the end of the biosynthesis, thus allowing extensive loss of deuterium label via keto–enol tautomerism.

2.8. Polivione is easily converted chemically into citromycetin (see problem 2.3), and this latter metabolite may thus be an artefact in this particular organism. A similar biogenesis to that of citromycetin seems reasonable, and the intermediate shown below has been proposed. This structure would possess the required $^{13}C-^{18}O$ linkages, and the appropriate number of ^{18}O atoms introduced during hydroxylation or oxidative cleavage.

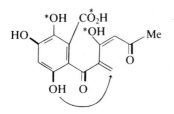

$^{18}O_2$ (*)

$Me^{13}C^{18}O_2^\ominus$ (—)

3.1. The intermediacy of the carbocation shown below seems reasonable, and equivalence of two of the carbon atoms derived from MVA is anticipated since these centres are connected by tautomerism.

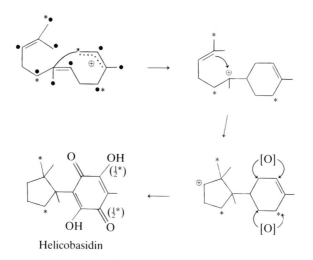

Helicobasidin

3.2. When [2-^{14}C, 5-^{3}H$_2$]-MVA was used (^{3}H:^{14}C ratio of 5.4), doubly labelled lactone was obtained with a ^{3}H:^{14}C ratio of 2.5. This implied that four moles of MVA were incorporated with a loss of four tritium atoms, and a calculated ^{3}H:^{14}C ratio of 2.7. If three moles of MVA had been utilized the final ratio would have been 3.6 since two tritium atoms out of six (i.e. one-third of the total) would have been lost. A reasonable diterpenoid precursor is shown below:

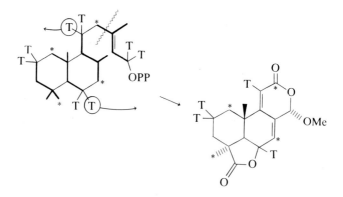

The ^{13}C-data are, of course, consistent with either a sequiterpenoid or diterpenoid pathway.

3.3. The presence of the cyclobutyl ring and the dimethylcyclopentyl ring are reminiscent of a proposed intermediate *en route* to illudin (see Fig. 3.15), and the route shown below thus seems reasonable. A more recent investigation by Donelly *et al.* on the biogenesis of the co-occurring metabolite fomajorin D, produced results that could also be accommodated by the proposed pathway.

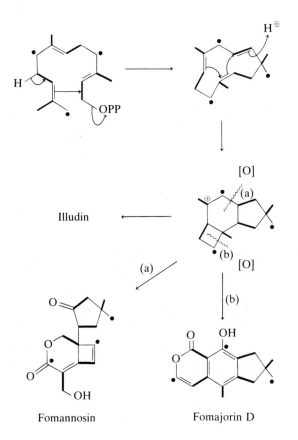

Illudin

Fomannosin

Fomajorin D

[Donelly, D. M. X., O'Reilly, J., Polonsky, J., and Sheridan, M. H. (1983). *Chem. Commun.*, 615]

3.4. The labelling pattern is, of course, consistent with assembly via a C_{15} precursor, but the position of the hydroxyls suggests that nerolidyl pyrophosphate may be the actual precursor. Indeed, Cane and coworkers, using cell-free enzymes

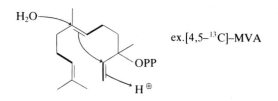

ex.[4,5–^{13}C]–MVA

from *Gibberella fujikuroi*, have successfully demonstrated the conversion of [1,2-^{13}C]-nerolidyl pyrophosphate into cyclonerodiol, and some of the pertinent n.m.r. data are given below.

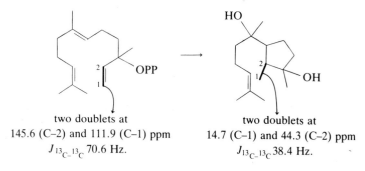

two doublets at
145.6 (C–2) and 111.9 (C–1) ppm
$J_{^{13}C–^{13}C}$ 70.6 Hz.

two doublets at
14.7 (C–1) and 44.3 (C–2) ppm
$J_{^{13}C–^{13}C}$ 38.4 Hz.

[Cane, D. E., Iyengar, R., and Shiao, M. (1981). *J. Am. Chem. Soc.*, **103**, 914]

3.5. There are two, most likely arrangements of isoprene units, I and II, but only I is compatible with the ^{13}C–^{13}C coupling data. In particular, note the coupling between the atoms marked with asterisks.

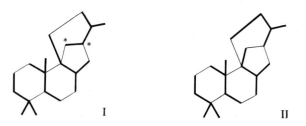

I

II

One potential mode of cyclization is shown below (and another is discussed in the paper cited); this is the most probable route, since incorporation of [3-^{13}C, 4-^{2}H$_2$]-MVA resulted in the observation of a ^{2}H–^{13}C coupling (J 18.5 Hz), and an isotope-induced upfield shift for the signal due to C-8. This is fully consistent with migration of a deuterium atom from C-9 to C-8.

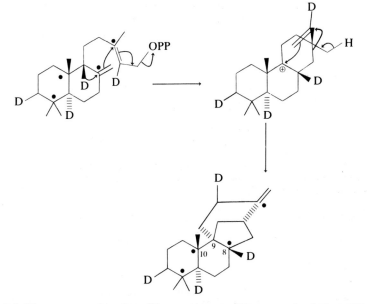

3.6. There are a number of possible arrangements of isoprene units: I, II, and III. Hanson and coworkers fed [1-¹³C]- and [1,2-¹³C]-acetates and obtained the

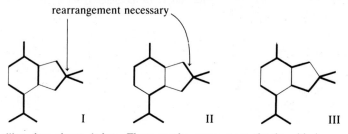

rearrangement necessary

I   II   III

labelling data shown below. These results agree most closely with isoprene arrangement III, and further efforts to clarify the biogenetic pathway have been executed.

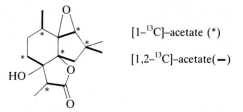

[1–¹³C]–acetate (*)

[1,2–¹³C]–acetate(—)

[Bradshaw, A. P. W., Hanson, J. R., and Sadler, I. H. (1982). *Chem. Commun.*, 292]

3.7. The metabolite appears to be a much modified diterpenoid, so geranylgeranyl pyrophosphate is the logical precursor.

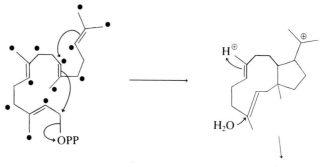

Fusicoccin

The sugar moiety also contains a part-structure derived from DMAPP, i.e.:

4.1. Chlorflavonin is a flavonoid, and provides a rare example of the formation of this type of secondary metabolite outside of the plant kingdom. Its biogenesis would seem to be straightforward:

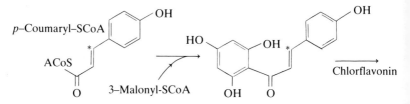

(*) ex.[3−^{14}C]-phenylalanine

However, recent work by Vining, McInnes, and their coworkers has shown that phenylalanine is degraded to an ArC$_1$ moiety prior to condensation with *four* C$_2$ units. This is another example of the common finding that different biosynthetic pathways operate in different classes of organisms, even if the final metabolites produced are similar in structure. The revised pathway is shown below.

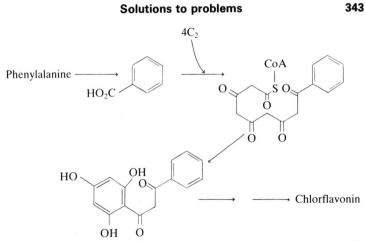

[Burns, M. K., Coffin, J. M., Kurobane, I., Vining, L. C. McInnes, A. G., Smith, D. G., and Walter, J. A. (1981). *Perkin Trans. 1.*, 1411]

4.2. [1,2-¹³C]-acetate was incorporated by stressed plants into eugenin and into 6-methoxymellein [227]. Although the latter produced a discrete pattern of ¹³C–¹³C couplings in its ¹³C-spectrum (five sets of discernible doublets), the spectrum of eugenin had couplings that suggested there were two labelling patterns. This means that free rotation must have been possible at some stage, and a reasonable biogenesis is shown below.

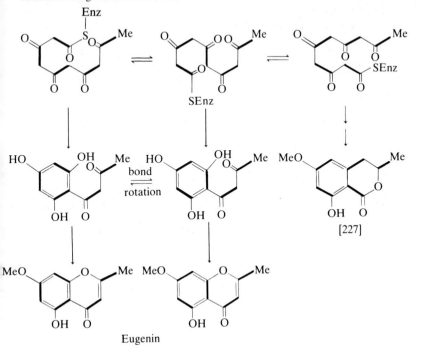

[227]

Eugenin

4.3. Psilotin has the appearance of a homologated cinnamic acid, and indeed [1-^{14}C, 2,3-^{13}C, 4'-^{3}H]-phenylalanine was incorporated without significant loss of any isotope; a reasonable pathway is shown below. Note that hydroxylation probably involves a 1,2-shift of the tritium atom—an example of the so-called NIH shift (see Chapter 2).

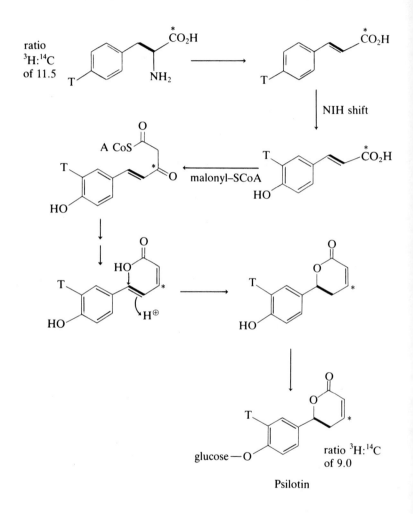

Psilotin

5.1. Pinidine is related in structure to coniine, and like this metabolite is derived via the polyketide pathway. The putative polyketide can be drawn in two ways, A and B, with the anticipated labelling patterns from [1-^{14}C]-malonate indicated on the structures. When [1-^{14}C]-malonate was used in a feeding experiment, and the resultant radio-labelled pinidine degraded, it was shown that although C-9

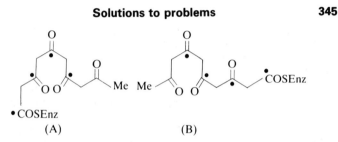

(A)   (B)

possessed 30% of the radioactivity, C-2 only had 9%. Hence, polyketide B (or its equivalent) seems most plausible.

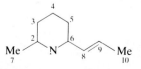

5.2. Stylopine would seem to arise via the normal pathway from reticuline, then it is proposed that ring cleavage and re-formation occur to yield chelidonine. This

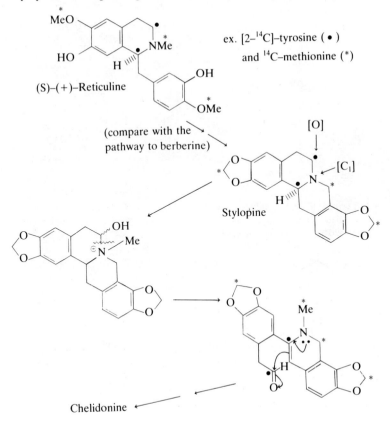

ex. [2–^{14}C]–tyrosine (•)

and ^{14}C–methionine (*)

(S)–(+)–Reticuline

(compare with the pathway to berberine)

[O]

[C₁]

Stylopine

Chelidonine

alkaloid is the most abundant member of the benzophenanthridine class, and possesses potent cytotoxic activity.

5.3. Anthramycin appears to be derived from anthranilic acid, tyrosine, and methionine, but the tyrosine moiety has suffered extensive degradation. By using [1-C,5′-³H]-tyrosine it was possible to show that the aromatic ring was cleaved as shown below (the tritium atom was retained in the aldehyde group), and after further cleavage and chain extension, the dihydropyrrole derivative is obtained.

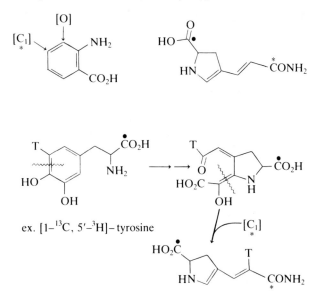

ex. [1–¹³C, 5′–³H]–tyrosine

5.4. Lysine is likely to produce cadaverine for incorporation, and this symmetrical intermediate should ensure that two discrete sets of ¹³C–¹³C couplings are observed. This was indeed the case.

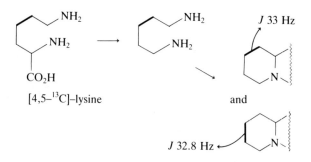

[4,5–¹³C]–lysine

The remainder of the molecule probably arises as shown below.

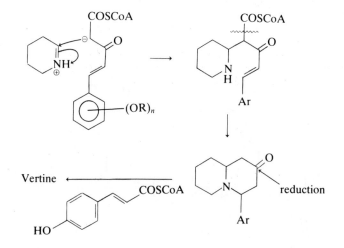

5.5. The probable amino acid precursors are anthranilic acid, phenylalanine, and methionine.

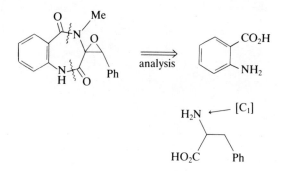

6.1. The metabolite appears to be derived from chalcone and two extra C_1 units ex. S-adenosyl methionine.

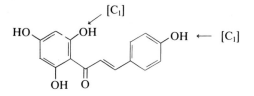

6.2. The metabolite has the appearance of a product of mixed metabolism of MVA and the polyketide pathways. Indeed, tracer studies with $[1,2\text{-}^{13}C]$-acetate have

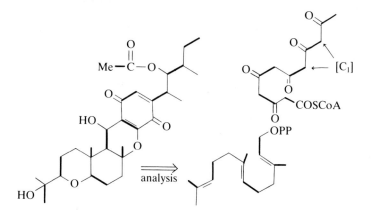

demonstrated the pattern shown above. Speculation concerning the later stages of the biogenesis is given in the paper, but the intermediate shown below is probably involved.

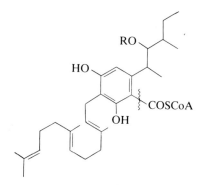

6.3. α-Cyclopiazonic acid appears to have an indole-C_2N moiety (ex. tryptophan), a five-carbon sub-unit from DMAPP, and four other carbon atoms of polyketide origin. The biogenesis has been shown (through use of the appropriate labelled precursors) to proceed as shown below. In addition, it is also known that the pro-(S)-3-hydrogen of tryptophan is lost in the final stages (the paper cited in the question contains details of the experiments used to establish how the final two bonds are made).

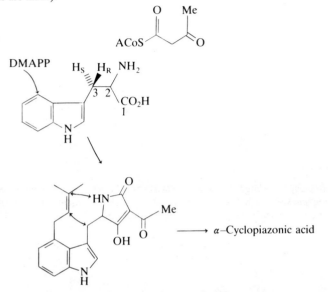

6.4. The route shown below is one of two possibilities. The alternative would involve reaction of malonyl-SCoA with dihydroferulyl-SCoA, and then a Claisen ester condensation with hexanoyl-SCoA.

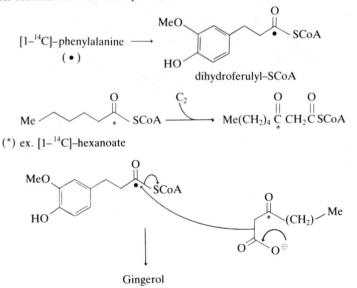

6.5. The right-hand portion of the molecule has the appearance of a polyketide metabolite, while MVA labels the left-hand portion. An analysis of the probable biogenetic precursors is given below.

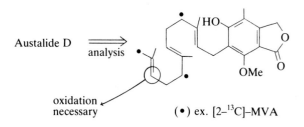

(•) ex. [2–^{13}C]–MVA

The phthalide is related in structure to mycophenolic acid, and probably arises via the sequence shown below. Speculation about the later (oxidative) stages of the biogenesis is given in the paper cited (in the question).

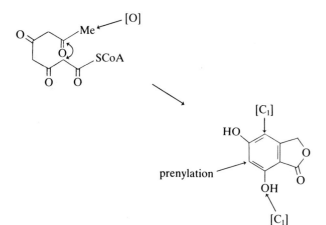

6.6. The metabolite is clearly a substituted naphthoquinone, which might reasonably arise via prenylation of lawsone. The biogenesis of compounds of this type was discussed in Fig. 6.9. ^{14}C-lawsone as its 2-isoprenyl ether was synthesized and shown to be an intermediate, and the mechanism below seems reasonable.

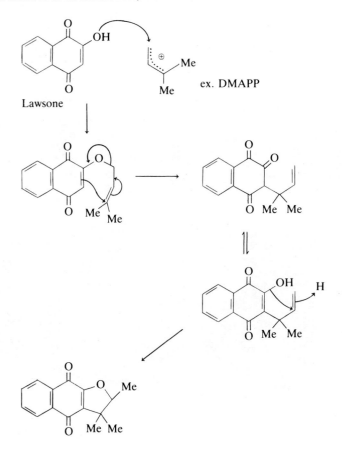

Suggested further reading

The following lists are not exhaustive, but the most important and easily accessible books and reviews have been selected.

Generally applicable texts

1. Crout, D. H. G. and Geissman, T. A. (1969). *Organic chemistry of secondary plant metabolism*. Freeman Cooper, San Francisco.
2. Carnduff, J., Murray, A. W., Nechvatal, A., and Tedder, J. M. (1972). *Basic Organic Chemistry*, Part IV, Wiley, Chichester.
3. Haslam, E. (ed.) (1979). *Comprehensive organic chemistry*. Vol. 5. Pergamon Press, Oxford.
4. Manitto, P. (1981). *Biosynthesis of natural products*. Ellis Horwood, Chichester.
5. Conn, E. E. (1981). *Secondary plant products*. Academic Press, New York.
6. Herbert, R. B. (1989). *Biosynthesis of secondary metabolites*. 2nd edn. Chapman & Hall, London.
7. Luckner, M. (1983). *Secondary metabolites in microorganisms, plants, and animals*. 2nd edn. Springer-Verlag, Berlin.
8. Torssell, K. B. G. (1983). *Natural product chemistry*. Wiley, Chichester.
9. Haslam, E. (1985). *Metabolites and metabolism*. Oxford University Press, Oxford.
10. Goodwin, T. W. and Mercer, E. I. (1983). *Introduction to plant biochemistry*. 2nd edn. Pergamon Press, Oxford.
11. Nakanishi, K., Goto, T., Ito, S., Natori, S., and Nozoe, S. (eds). Vols 1 and 2 (1974), Academic Press, New York; Vol. 3 (1983), Oxford University Press, Oxford.
12. Thomson, R. H. (ed.) (1985). *The chemistry of natural products*. Blackie, Glasgow and London.
13. Geissman, T. A. (ed.). *Biosynthesis*, Vols 1–3, Royal Society of Chemistry, London; and Bu'Lock, J. D. (ed.), Vols 4–7.
14. Pattenden, G. (ed.). *Natural Product Reports*. Vols 1–7, and further volumes forthcoming, Royal Society of Chemistry, London.

   This journal was first published in February 1984, and has appeared at the rate of six issues per year ever since. It replaced the Specialist Periodical Reports entitled *Aliphatic and related natural product chemistry*, *The Alkaloids*, *Terpenoids and Steroids*, and the above mentioned *Biosynthesis*. It is presently the best source of current information on all aspects of natural products.

15. Elvin-Lewis, M. P. F. and Lewis, W. H. (1977). *Medical botany: plants affecting man's health*. Wiley–Interscience, New York.

Finally, a large number of books and review articles have appeared which deal with marine natural products, and the following are of note:

Scheuer, P. J. (ed.) (1985). *Tetrahedron* **41**, 979–1108.
Naylor, S. (1984). *Chemistry in Britain* **20**, 118.
Faulkner, D. J. (1990). *Natural Product Reports* **7**, 269.

Chapter 1

1. *Primary metabolic pathways, enzymes, and cofactors*:
   Mahler, H. R. and Cordes, E. H. (1971). *Biological chemistry*. Harper & Row, New York.
   Stryer, L. (1988). *Biochemistry*. 3rd edn. Freeman, San Francisco.
   Metzler, D. E. (1977). *Biochemistry*. Academic Press, New York.
   Lehninger, A. L. (1983). *Principles of biochemistry*. 2nd edn. Worth, New York.
   Staunton, J. (1978). *Primary metabolism* (OCS). Oxford University Press, Oxford.
   Suckling, K. E. and Suckling, C. J. (1980). *Biological chemistry*. Cambridge University Press, Cambridge.
2. *Phytochemical methods*:
   Harborne, J. B. (1984). *Phytochemical methods—a guide to techniques of plant analysis*. 2nd edn. Chapman & Hall, London.
3. *Biological methods for studying the biosynthesis of natural products*:
   Hutchinson, C. R. (1986). *Natural Product Reports* **3**, 133.
4. *Chemotaxonomy*:
   Harborne, J. B. and Turner, B. L. (1984). *Plant chemosystematics*. Academic Press, London.

Chapter 2

1. Packter, N. M. (1973). *Biosynthesis of acetate-derived compounds*. Wiley, London.
2. Edwards, J. M. and Weiss, U. (1976). *The biosynthesis of aromatic compounds*. Wiley, Chichester.
3. *Biosynthesis of fatty acids:*
   Chang S-I and Hammes, G. G. (1990). *Acc. Chem. Res.* **23**, 363.
4. *Fatty acids and glycerides (general chemistry and biological properties)*:
   Lie, M. S. F. and Tie, K. *Natural Products Reports* **6**, 231.
5. *Polyacetylenes*:
   Bohlmann, F., Burkhardt, T., and Zdero, C. (1973). *Naturally occurring acetylenes*. Academic Press, London.
   Hansen, L. and Boll, P. M. (1986). *Phytochemistry* **25**, 285.
6. *Prostanoids and leukotrienes*:
   Lai, S. M. F. and Manley, P. W. (1984). *Natural Products Reports* **1**, 409.
7. *Polyketides*:
   Turner, W. B. (1971). *Fungal metabolites*. Academic Press, London.
   Aldridge, D. C. and Turner, W. B. (1983). *Fungal metabolites II*. Academic Press, London.
   Simpson, T. J. (1987). *Natural Product Reports*.
8. *N.m.r. techniques*:
   Simpson, T. J. (1975). *Chem. Soc. Rev.* **4**, 497
   Garson, M. J. and Staunton, J. (1979). *Chem. Soc. Rèv.* **8**, 539.

Verderas, J. C. (1982). *Canad. J. Chem.* **60**, 1637.

Benn, R. and Günther, H. (1983). *Angew. Chem.* **22**, 350.

Scott, A. I. (ed.) (1983). *Tetrahedron* **39**, 3441–591.

9. *Mycotoxins*:
   Burchard, F. (1984). *Angew. Chem.* **23**, 493.
   Steyn, P. S. (ed.) (1980). *Biosynthesis of mycotoxins.* Academic Press, New York.

10. *Tetracyclines*:
    Hutchinson, C. R. (1981). *Antibiotics* (ed. J. W. Corcoran), Vol. 4, p. 1. Springer-Verlag, New York.

11. *Macrocyclic antibiotics*:
    Hutchinson, C. R. (1983). *Acc. chem. Res.* **16**, 7.
    Jones, R. C. F. (1984). *Natural Product Reports* **1**, 87.

12. *Avermectins and milbemycins*:
    Davies, H. G. and Green, R. H. (1986). *Natural Product Reports* **3**, 87.

Chapter 3

1. *Isoprene rule and biogenetic isoprene rule*:
   Ruzicka, L. (1953). *Experientia* **9**, 357;
   —— (1970). *Ann. Rev. Biochem.* **42**, 1.

2. *Biogenetic relationships and biosynthesis of isoprenoids in general*:
   Porter, J. W. and Spurgeon, S. L. (eds). *Biosynthesis of isoprenoid compounds*, Vol. 1 (1981), and Vol. 2 (1984).

3. *Biosynthesis of C_5 to C_{20} isoprenoids*:
   Beale, M. H. (1990). *Natural Products Reports* **7**, 25.

4. *Biosynthesis of triterpenoids and steroids*:
   Harrison, D. M. (1990). *Natural Product Reports* **7**, 459.
   Caspi, E. (1986). *Tetrahedron* **42**, 3.

5. *Biosynthesis of steroids*:
   Schroepker, G. J. (1981 and 1982). *Ann. Rev. Biochem.* **50**, 585, and **51**, 555;
   Law, J. H. and Rilling, H. C. (eds) (1985). *Methods in enzymology: Steroids and isoprenoids*, parts A and B. Academic Press, New York.

6. *Biosynthesis of carotenoids and higher terpenoids*:
   Harrison, D. M. (1986). *Natural Product Reports* **3**, 205.

7. *Many aspects of isoprenoid chemistry, biosynthesis, and biological function*:
   [Various authors] (1983). *Biochem. Soc. Trans.* **11**, 473–611.

8. *Good general text on chemistry and biosynthesis of all isoprenoids except steroids*:
   Newman, A. A. (ed.) (1972). *Chemistry of terpenes and terpenoids.* Academic Press, London.

Chapter 4

1. Haslam, E. (1975). *The shikimate pathway.* Butterworths, London.

2. *Biosynthesis of shikimate metabolites*:
   Dewick, P. M. (1990). *Natural Product Reports* **7**, 165.

3. *Enzymology of the shikimate pathway*:
   Walsh, C. T., Liu, J., Rusnak, F., and Sakaitani, M. (1990). *Chem. Rev.* **90**, 1105.

4. Murrey, R. D. H., Méndez, J. and Brown, S. A. (1982). *The natural coumarins: occurrence, chemistry, and biochemistry.*
5. *Lignans*:
MacRae, W. D. and Towers, G. H. N. (1984). *Phytochemistry* **23**, 1207.
Whiting, D. A. (1990). *Natural Product Reports* **7**, 349.

Chapter 5

1. Cordell, G. A. (1981). *Introduction to the alkaloids: a biogenetic approach.* Wiley, New York.
2. *Alkaloids.* Vols 1–26 (1950–86); Manske, R. H. F. and Holmes, H. L. (eds) Vols 1–5; Manske (Vols 6–16); Manske and Rodrigo, R. (Vols 17–20); Brossi, A. (Vols 21–26). Academic Press, New York.
3. Pelletier, S. W. *Alkaloids; chemistry and biological perspectives*, Vol. 1 (1983), Vol. 2 (1985), and Vol. 3 (1985). Wiley, New York.
4. *Biosynthesis of alkaloids*:
Herbert, R. B. (1990). *Natural Product Reports* **7**, 105.
5. *Novel amino acids*:
Wagner, I. and Musso, H. (1983). *Angew. Chem.* **22**, 816.
6. *Chemistry and biology of beta-lactam antibiotics*:
Morrin, R. B. and Gorman, M. (eds) (1982). *Chemistry and biology of beta-lactam antibiotics*. Vols 1, 2, and 3. Academic Press, New York.
7. *Biosynthesis of beta-lactam antibiotics*:
Baldwin, J. E. and Bradley, M. (1990). *Chem. Rev.* **90**, 1079.
8. *Biochemistry of alkaloids*:
Luckner, M., Mothes, K. and Schutte, H. R. (1985). *Biochemistry of the alkaloids.* V.C.H., Weinheim (available via the Royal Society of Chemistry).

Chapter 6

1. *Cannabinoids*:
Mechoulam, R., Burstein, S. and McCallum, N. K. (1976). *Chem. Rev.* **76**, 75.
2. *Biosynthesis of isoprenoid quinones*:
Pennock, J. F. (1983). *Biochem. Soc. Trans.* **11**, 504.
3. *Flavonoids*:
Harborne, J. B., Mabry, T. J., and Mabry, H. (1982). *The flavonoids*, 2nd edn. Chapman & Hall, London.
4. *Indole alkaloid biosynthesis*:
Scott, A. I. (1981). *Heterocycles* **15**, 1257.;
Rahman, A-U, and Basha, A. (1983). *Biosynthesis of indole alkaloids.* Oxford University Press, New York.
5. *Ergot alkaloids*:
Floss, H. G. (1976). *Tetrahedron* **22**, 873.

Chapter 7

1. *Three excellent general texts are available*:
Simeone, J. B. and Sondheimer, E. (eds) (1970). *Chemical ecology.* Academic Press, London.
Harborne, J. B. (1971). *Phytochemical ecology.* Academic Press, London;
——(1988). *Introduction to ecological biochemistry.* 3rd edition. Academic Press, London.

2. *Insect pheromones*:
   Baker, R. and Herbert, R. B. (1984). *Natural Product Reports* **1**, 299.
3. *Animal communication*:
   Albone, J. (1984). *Mammalian semiochemicals: investigation of the chemical signals*. Wiley, New York.
   —— (1982). *Tetrahedron* **38**, 1853–970.
4. *Allelopathic agents*:
   Putnam. A. R. (1983). *Chem. & Engng. News* 34.
   Rice, E. L. (1984). *Allelopathy*, 2nd edn. Academic Press, Orlando.
5. *Phytoalexins*:
   Bailey, J. A. and Mansfield, J. N. (eds) (1982). *Phytoalexins*. Blackie, Glasgow.
   Brooks, C. J. W. and Watson, D. G. (1985). *Natural Product Reports* **2**, 427.
6. *Irritant and defence substances in higher plants*:
   Meinwald, J. (1982). *Tetrahedron* **38**, 1853–970
7. *Chemicals from the glands of ants*:
   Attygalle, A. B. and Morgan, E. D. (1984). *Chem. Soc. Rev.* **13**, 245.
8. *Recent advances in chemical ecology*:
   Harborne, J. B. (1989). *Natural Product Reports* **6**, 85.

Selected research references and specialized reviews

The following references are taken from the primary research literature, and include the latest (up to Jan. 1991) or the most important papers pertaining to secondary metabolites mentioned in the text. Numbers in square brackets refer to the structures given in the text. The best source of information about all aspects of natural products is *Natural Product Reports*, Royal Society of Chemistry.

Chapter 2

 1. *Mechanism of action of biotin* [7]:
    Knowles, J. R. and O'Keefe, S. J. (1986). *J. Am. Chem. Soc.* **108**, 328.
 2. *Stereochemistry of fatty acid biosynthesis*:
    Cornforth, J. W., French, S. J., Gray, R. T., Kelstrup, E., Sedgwick, B., and Willadsen, P. (1977). *Eur. J. Biochem.* **75**, 465–95.
    French, S. J., Morris, C., and Sedgwick, B. (1978). *Chem. Commun.*, 193.
    McInnes, A. G., Walter, J. A., and Wright, J. L. C. (1983). *Tetrahedron* **39**, 3515.
 3. *Anaerobic route to unsaturated fatty acids*:
    Klassen, J. B., Habib, A., and Schwab, J. M. (1986). *J. Am. Chem. Soc.*, **108**, 5304.
 4. *Biosynthesis of fatty acids by* Bacillus acidocaldarius [9]:
    Bu'Lock, J. D., Gambacorta, A. and DeRosa, M. (1974). *Phytochem.* **13**, 905.
 5. *Biosynthesis of matricaria ester* [29]:
    Bohlmann, F. and Burkhardt, T. (1972). *Chem. Ber.* **105**, 521.
 6. *Biosynthesis of lactobacillic acid* [31]:
    Buist, P. H. and Findlay, J. M. (1985). *Canad. J. Chem.* **63**, 972.
 7. *Biosynthesis of 6-methylsalicylic acid* [41]:
    Abell, C. and Staunton, J. (1984). *Chem. Commun.*, 1005.
    —— and —— (1981). *Chem. Commun.*, 856.
 8. *Biosynthesis of griseofulvin* [42]:
    Lane, M. P., Nakashima, T. T., and Vederas, J. C. (1982). *J. Am. Chem. Soc.* **104**, 913.
 9. *Biosynthesis of dihydrolactumcidin*:
    Sato, T., Seto, H., and Yonehara, H. (1973). *J. Am. Chem. Soc.* **95**, 8461.
10. *Biosynthesis of asperlin* [44]:
    Hamasaki, T., Johnson, L., Tanabe, M., and Thomas, D. (1971). *J. Am. Chem. Soc.* **93**, 273.
11. *Biomimetic oxidative cleavage of aromatic rings*:
    Pandell, A. J. (1983). *J. Org. Chem.* **48**, 3908.
12. *Biosynthesis of penicillic acid* [49]:
    Elvidge, J. A., Jaiswal, D. K., Jones, J. R., and Thomas, R. (1977). *Perkin Trans. I*, 1080.
13. *Biosynthesis of patulin* [50]:
    Chou, D. T-W., and Ganem, B. (1980). *J. Am. Chem. Soc.* **102**, 7987.

Gaucher, G. M., Sekiguchi, J., Shimamoto, T. and Yamada, Y. (1983). *Appl. environ. Micro.* **45**, 1939.

14 *Biosynthesis of citrinin* [51]:

Barber, J., Carter, R. H., Garson, M. J., and Staunton, J. (1981). *Perkin Trans. 1*, 2577.

Ebizuka, Y., Iitaka, Y., Ishikawa, Y., Kitagawa, S., Kobayashi, T., Noguchi, H., Sankawa, U., Seto, H., and Yamamoto, Y. (1983). *Tetrahedron* **39**, 3583

15. *Biosynthesis of sepedonin* [53]:

Johnson, L., McInnes, A. G., Smith, D. G., and Vining, L. C. (1971). *Chem. Commun.*, 325.

16. *Biosynthesis of stipitatonic acid* [54]:

Scott, A. I. (1971). *Pure appl. Chem.* **5**, 34.

17. *Biosynthesis of variotin* [57]:

Tanabe, M. and Seto, H. (1970). *Biochim. biophys. Acta.* **208**, 151.

18. *Biosynthesis of rubrofusarin* [58]:

Leeper, F. J. and Staunton, J. (1984). *Perkin Trans I*, 2919.

19. *Biomimetic synthesis of polyketides*:

Griffin, D. A., Leeper, F. J., and Staunton, J. (1984). *Perkin Trans. 1*, 1035, 1053.

Abell, C., Bush, B. D. and Staunton, J. (1986). *Chem. Commun.*, 15.

20. *Biosynthesis of tajixanthone*:

Bardshiri, E., McIntyre, C. R., Moore, R. N., Simpson, T. J., Trimble, L. A., and Vederas, J. C. (1984). *Chem. Commun.*, 1404.

21. *Biosynthesis of aflatoxins*:

Townsend, C. A. (1986). *Pure appl. Chem.* **58**, 227.

22. *Biosynthesis of secalonic acids* [64]:

Kurobane, I., McInnes, A. G., Walters, J. A., Wright, J. L. C., and Vining, L. C. (1978). *Tetrahedron Lett.*, 1379.

23. *Biosynthesis of tetracyclines*:

Thomas, R. and Williams, D. J. *Chem. Commun.* (1984) 443, and (1985) 802.

24. *Biosynthesis of sclerin* [76]:

Barber, J., Garson, M. J., and Staunton, J. (1981). *Perkin Trans. 1*, 2584.

25. *Biosynthesis of erythromycins*:

Cane, D. E., Hasler, H., Liang, T-C., and Taylor, P. B. (1983). *Tetrahedron* **39**, 3449.

26. *Biosynthesis of ansamycins*:

Becker, A. M., Herlt, A. J., Hilton, G. L., Kibby, J. J., and Rickards, R. W. (1983). *J. Antibiotics* **36**, 1323.

27. *Biosynthesis of lasalocid A* [82]:

Hutchinson, C. R., McInnes, A. G., Nakashima, T. T., Sherman, M. M., Vederas, J. C., and Walter, J. A. (1981). *J. Am. Chem. Soc.* **103**, 5933, 5956.

28. *Biosynthesis of milbemycins and avermectins*:

Iwasaki, S., Kobayashi, H., Mishima, H., Ono, M., Okuda, S., Takiguchi, Y., and Terao, M. (1983). *J. Antibiotics* **36**, 991.

29. *Biosynthesis of the tetronic acids* [84]:

Holker, J. S. E., Moore, R. N., O'Brien, E., and Vederas, J. C. (1983). *Chem. Commun.*, 192.

Chapter 3

1. *Mechanism of prenyl transferase reaction*:
Gurria, G. M., Le, A. T., Mash, E. A., Poulter, C. D., and Wiggins, P. L. (1981). *J. Am. Chem. Soc.* **103**, 3926, 3927.

2. *Enzymic aspects of monoterpene and sesquiterpene biosynthesis*:
Cori, O. M. (1983). *Phytochemistry* **22**, 331.
Cane, D. E. (1983). *Biochem. Soc. Trans.* **11**, 510

3. *Cyclisation of allylic phosphates to yield mono- and sesquiterpenes (involvement of nerolidyl pyrophosphate)*:
Cane, D. E., Chang, C., Croteau, R., Saito, A., and Shaskus, J. (1984). *J. Am. Chem. Soc.* **106**, 1142.
Cane, D. E. (1985). *Acc. Chem. Res.* **18**, 220.
Davisson, V. J., Neal, T. R., and Poulter, C. D. (1985). *J. Am. Chem. Soc.* **107**, 5277.

4. *Biosynthesis of thujone* [91]:
Banthorpe, D. V., Mann, J., and Turnbull, K. W. (1970). *J. Chem. Soc.* (C), 2689.

5. *Biosynthesis of pinenes*:
Banthorpe, D. V., Ekundayo, O, and Njar, V. C. O. (1984). *Phytochemistry* **23**, 291.

6. *Biosynthesis of irregular terpenes*:
Poulter, C. D. (1990). *Acc. Chem. Research* **23**, 70.

7. *Pyrethrins*:
Elliott, M. and Janes, N. F. (1978). *Chem. Soc. Rev.* **7**, 473.
[Various authors] (1985). In *Recent advances in the chemistry of insect control* (ed. N. F. Janes). Royal Society of Chemistry, London.

8. *Model studies for the biosynthesis of irregular monoterpenes*:
King, C-H, R., and Poulter, C. D. (1982). *J. Am. Chem. Soc.* **104**, 1413.

9. *Biosynthesis of iridoids*:
Tietze, L-F. (1983). *Angew. Chem.* **22**, 828
Iida, A., Inouye, H., Kanomi, S., Uesato, S., and Zenk, M. H. (1986). *Phytochemistry* **25**, 839.

10. *Biosynthesis of bisabolene* [125] *and paniculide*:
Anastasis, P., Freer, I., Gilmore, C., Mackie, H., Overton, K., and Picken, D. (1984). *Canad. J. Chem.* **62**, 2079.

11. *Biosynthesis of ipomearone* [131]:
Lee, J., Mizukawa, K., Nakanishi, K., Schneider, J., and Yoshihara, K. (1984). *Chem. Commun.*, 372.

12. *Biosynthesis of juvenile hormones* [132]:
Feyereisen, R., Ruegg, R. P., and Tobe, S. S. (1984). *Insect Biochem.* **14**, 657.

13. *Biosynthesis of ovalicin* [133]:
Buchwald, S. L., and Cane, D. E. (1977). *J. Am. Chem. Soc.* **99**, 6132.

14. *Biosynthesis of trichodiene* [135]:
Cane, D. E. and Ha, H-J. (1986). *J. Am. Chem. Soc.* **108**, 3097.

15. *Biosynthesis of illudins*:
Avent, A. G., Hanson, J. R., and Yeoh, B. L. (1985). *J. Chem. Res.* (S), 396.

16. *Biosynthesis of rosenonolactone* [165]:
Cane, D. E. and Murthy, P. P. (1977). *J. Am. Chem. Soc.* **99**, 8327.

17. *Biosynthesis of virescenol*:
Cagnoli-Bellavita, N., Ceccherelli, P., Lukacs, G., and Polonsky, J. (1975). *Tetrahedron Lett.*, 481.

18. *Biosynthesis of gibberellins*:
Hanson, J. R. (1983). *Biochem. Soc. Trans.* **11**, 522.
Crozier, A., Graebe, J. A., Schwenen, L., and Turnbull, C., G. (1986). *Phytochemistry* **25**, 97.

19. *Biosynthesis of ophiobolanes*:
(Summarized by) Hanson, J. R. (1986). *Natural Product Reports* **3**, 128.

20. *Biosynthesis of presqualene pyrophosphate* [172] *and models for squalene* [171] *biosynthesis*:
Leopold, E. J. and Van Tamelen, E. E. (1985). *Tetrahedron Lett.*, 3303.

21. *The 2,3-oxidosqualene to lanosterol conversion*:
Van Tamelen, E. E. (1982). *J. Am. Chem. Soc.* **104**, 6480.

22. *Biomimetic steroid synthesis*:
Van Tamelen, E. E. (1981). *Pure appl. Chem.* **53**, 1259.

23. *Lanosterol to cholesterol pathways*:
(Summarized by) Harrison, D. M. (1985). *Natural Product Reports* **2**, 534.

24. *Biosynthesis of triterpenoids and steroids in higher plants, algi, and fungi*:
(Summarized by) Harrison, D. M. (1985). *Natural Product Reports* **2**, 543.

25. *Biosynthesis of tetrahymanol* [181]:
Aberhart, D. J. and Caspi, E. (1979). *J. Am. Chem. Soc.* **101**, 1013.

26. *Alkylation of phytosterol side-chains*:
Seo, S., Takeda, K., Yomori, A., and Yoshimura, Y. (1983). *J. Am. Chem. Soc.* **105**, 6343.

27. *Degradation of steroid side-chains by insects*:
(Summarized by) Harrison, D. M. (1985). *Natural Product Reports* **2**, 551.

28. *Biosynthesis of ecdysones and other sterol metabolism in insects*:
Ikewawa, N. (1983). *Experientia* **39**, 466.

29. *Biosynthesis of diosgenin* [185]:
Seo, S., Tori, K., Uomori, A., and Yoshimura, Y. (1981). *Chem. Commun.*, 895

30. *Biosynthesis of steroidal hormones*:
(Summarized by) Harrison, D. M. (1985). *Natural Product Reports* **2**, 537

31. *Biosynthesis of estrogens*:
Akhtar, M., Stevenson, D. E., and Wright, J. N. (1985). *Chem. Commun.*, 1078.

32. *Biosynthesis of bile acids*:
(Summarized by) Harrison, D. M. (1985). *Natural Product Reports* **2**, 542.

33. *The vitamins D* [89]:
DeLuca, H. F. and Schnoes, H. K. (1983). *Ann. Rev. Biochem.* **52**, 411.

34. *Biosynthesis of carotenoids*:
Goodwin, T. W. (1983). *Biochem. Soc. Trans.* **11**, 473.

35. *Photoreceptor pigments* (*vitamin A section*):
(Summarized by) Britton, G. (1985). *Natural Product Reports* **2**, 375.

36. *Trisporic acids* [207]:
Griffin, D., Horgan, R., Neill, S. J., and Walton, D. C. (1983). *Biochem. Soc. Trans.* **11**, 553

37. *Biosynthesis of abscisic acid* [208]:
(Summarized by) Banthorpe, D. V. and Branch, S. A. (1985). *Natural Product Reports* **2**, 518.

Chapter 4

1. *Mechanism of the process catalysed by chorismate mutase*:
Copely, S. D. and Knowles, J. R. (1985). *J. Am. Chem. Soc.* **107**, 5306
Bartlett, P. A. and Johnson, C. R. (1985). *J. Am. Chem. Soc.* **107**, 7792.

2. *Role of phenylalanine ammonia-lyase*:
Jones, D. H. (1984). *Phytochemistry* **23**, 1349.

3. *Cinnamic acid esters, lignins, lignans, and coumarins*:
(Summarized by) Dewick, P. B. (1985). *Natural Product Reports* **2**, 499

4. *Biosynthesis of 6-methoxymellein* [227]:
Stoessl, A. and Stothers, J. B. (1978). *Canad. J. Bot.* **56**, 2589.

5. *Biosynthesis of allylphenols*:
Canonica, L., Gramatica, P., Manitto, P. and Monti, D. (1978). *Chem. Commun.*, 1073.

6. *Biosynthesis of cocaine* [267]:
Leete, E. and Kim, S. H. (1988). *J. Am. Chem. Soc.* **110**, 2976.

7. *Urinary lignans*:
Setchell, K. D. R., Lawson, A. M., Conway, E., Taylor, N. F., Kirk, D. N., Cooley, G., Farrant, R. D., Wynn, S., and Axelson, M. (1981). *Biochem. J.* **197**, 447.

8. *Comparison of biosynthetic pathways to salicylic acid* [246] *and 6-methyl-salicylic acid* [41]:
Marshall, B. J. and Ratledge, C. (1972). *Biochem. biophys. Acta* **264**, 106

9. *Biosynthesis of chloramphenicol* [251]:
Gottlieb, D., Munro, M. H. G., Rinehart, K. L., and Taniguchi, M. (1975). *Tetrahedron Lett.*, 2659

Chapter 5

1. *Biosynthesis of* (−)-*sedamine*:
Gupta, R. N., and Spenser, I. D. (1970). *Phytochemistry* **9**, 2329.

2. *Hypothetical scheme for the biosynthesis of piperidine alkaloids*:
Leistner, E., and Spenser, I. D. (1973). *J. Am. Chem. Soc.* **95**, 4715.

3. *Pyrrolidine, piperidine, and pyridine alkaloids*:
Pinder, A. R. (1986). *Natural Products Reports* **3**, 171.

4. *Tropane alkaloids*:
Fodor, G. and Dharanipragada, R. (1986). *Natural Products Reports* **3**, 181.

5. *Biosynthesis of scopolamine* [264] *and nicotine* [265]:
Leete, E. and McDonnell, J. A. (1981). *J. Am. Chem. Soc.* **103**, 658.

6. *Biosynthesis of cocaine* [267]:
Leete, E. (1983). *J. Am. Chem. Soc.* **105**, 6727.

7. *Biosynthesis of tropic acid* [269]:
Haslam, E., Opie, C. T., and Platt, R. V. (1984). *Phytochemistry* **23**, 2211.
Leete, E. (1984). J. Am. Chem. Soc. **106**, 7271.

8. *Enzymes of pyrrolidine and piperidine alkaloid biosynthesis*:
Richards, J. C. and Spenser, I. D. (1983). *Tetrahedron* **39**, 3549.
Battersby, A. R., Staunton, J., and Tippett, J. (1982). *Perkin Trans. I*, 455.

9. *Biosynthesis of coniine* [275]:
Roberts, M. F. (1978). *Phytochemistry* **17**, 107.

10. *Pyrrolizidine alkaloids*:
Robins, D. J. (1986). *Natural Products Reports* **3**, 297.

11. *Biosynthesis of retronecine* [281]:
Rana, J. and Robins, D. J. (1986). *Perkin Trans. I*, 983.

12. *Stereochemical aspects of the biosynthesis of pyrrolizine and quinolizidine alkaloids*:
Spenser, I. D. (1985). *Pure. appl. Chem.* **57**, 453.

13. *Biosynthesis of sparteine* [286] *and lupinine* [285]:
Golebiewski, W. M. and Spenser, I. D. (1984). *J. Am. Chem. Soc.* **106**, 7925.
Rana, J. and Robins, D. J. (1986). *Perkin Trans. I*, 1133.

14. *Biosynthesis of nicotine* [265]:
Leete, E. and McDonell, J. A. (1981). *J. Am. Chem. Soc.* **103**, 658.

15. *Biosynthesis of anatabine* [290]:
Leete, E. and Mueller, M. E. (1982). *J. Am. Chem. Soc.* **104**, 6440.

16. *Biosynthesis of ricinine* [295]:
Robinson, T. (1978). *Phytochemistry* **17**, 1903.

17. *Biosynthesis of gliotoxin* [303]:
Johns, N. and Kirby, G. W. (1985). *Perkin Trans. I*, 1487.

18. *Biosynthesis of betanin* [304]:
Dreiding, A. S., Dunkelblum, E., Fischer, N., and Miller, H. E. (1972). *Helv. chim. Acta.* **55**, 642.

19. *Isoquinoline alkaloid biosynthesis*:
(Summarized by) Lundstrom, J. (1983). In *The alkaloids* (ed. A Brossi), Vol. XXI, p. 312.

20. *Biosynthesis of benzylisoquinoline alkaloids*:
(Summarized by) Herbert R. B. (1986). *Natural Product Reports* **3**, 185.

21. *Enzymes of berberine* [314] *biosynthesis*:
Rueffer, M. and Zenk, M. H. (1985). *Tetrahedron Lett.*, 201.

22. *Biosynthesis of morphine* [315]:
Horn, J. S., Paul, A. G., and Rapoport, H. (1978). *J. Am. Chem. Soc.* **100**, 1895.

23. *Enkephalins* [321] *and other peptide opiates*:
Chapouthier., G. and Rossier, J. (1982). *Endeavour* **6**, 168.

24. *Biomimetic synthesis of thebaine* [323]:
Mami, I. S. and Schwartz, M. A. (1975). *J. Am. Chem. Soc.* **97**, 1239.

25. *Biosynthesis of erythrina alkaloids*:
(Summarized by) Dyke, S. F. and Quessy, S. N. (1981). In *The alkaloids*, (eds R. H. F. Manske and R. G. A. Rodrigo), Vol. XVIII, p. 51.

26. *Amaryllidaceae alkaloid biosynthesis*:
(Summarized by) Fuganti, C. (1975). In *The alkaloids*, (ed. R. H. F. Manske), Vol. XV, p. 145

27. *Biosynthesis of aporphine alkaloids*:
(Summarized by) Kametani, T. and Honda, T. (1985). In *The alkaloids*, (ed. A. Brossi), Vol. XXIV, p. 205.

28. *Biosynthesis of colchicine* [338]:
Herbert, R. B. and Knagg, E. (1986). *Tetrahedron Lett.*, 1099.
Battersby, A. R., McDonald, E., and Stachulski, A. V. (1983). *Perkin Trans. 1*, 3053.

29. *Biosynthesis of acridone alkaloids*:
(Summarized by) Herbert R. B. (1985). *Natural Product Reports* **2**, 174.

30. *Isolation of ajoene*:
Ahmad, S., Apitz-Castro, R., Block, E., Cruz, M. R., Crecely, R. W., and Jain, M. K. (1984). *J. Am. Chem. Soc.* **106**, 8295.

31. *Biosynthesis of penicillins and cephalosporins*:
(Summarized by) Robinson, J. A. and Gani, D. (1985). *Natural Product Reports* **2**, 293.
Baxter, R. L., McGregor, C. J., Scott, A. I., and Thomson, G. A. (1985). *Perkin Trans. 1*, 369.
Adlington, R. M., Baldwin, J. E., Basak, A., Flitsch, S. L., Peturson, S., Turner, N. J., and Ting, H-H. (1986). *Chem. Commun.*, 273, 975.

32. *Clavulanic acid biosynthesis*:
Ho, M-F., Mao, S-S., and Townsend, C. A. (1986). *Chem. Commun.*, 638.

27. *Biosynthesis of aporphine alkaloids*:
(Summarized by) Kametani, T. and Honda, T. (1985). In *The alkaloids*, (ed. A. Brossi), Vol. XXIV, p. 205.

Chapter 6

1. *Biosynthesis of mycophenolic acid* [366]:
Arogozzini, F., Colombo, L., Gennari, C., Gualandris, R., Potenza, D., and Scolastico, C. (1982). *Perkin Trans. 1*, 365.

2. *Biosynthesis of siccanochromenes*:
Nozoe, S. and Suzuki, K. T. (1972). *Chem. Commun.*, 1166.

3. *Biosynthesis of cannabinoids*:
Crombie, L. (1986). *Pure appl. Chem.* **58**, 693.
Jima, M. K. A. and Piraux, M. (1982). *Phytochemistry* **21**, 67.

4. *Biosynthesis of isoprenoid quinones*:
Pennock, J. F. (1983). *Biochem. Soc. Trans.* **11**, 504.

5. *Biosynthesis of furanocoumarins*:
Caporale, G., Dall'Acqua, F., and Innocenti, G. (1983). *Phytochemistry* **22**, 2207.

6. *Biosynthesis of furanoquinolines*:
Grundon, M. F., Harrison, D. M., and Spyropoulos, C. G. (1975). *Perkin Trans. 1*, 302.

7. *Biosynthesis of alizarin* [60]:
Leistner, E. (1973). *Phytochemistry* **12**, 337.

8. *Biosynthesis of flavonoids*:
(Summarized by) Dewick, P. M. (1985). *Natural Product Reports* **2**, 502.

9. *Rotenoids*:
Crombie, L. (1984). *Natural Product Reports* **1**, 3.

10. *Biosynthesis of proanthocyanidins*:
Foo, L. Y. (1986). *Chem. Commun.*, 675.
11. *Biosynthesis of pisatin* [404] *and phaseollin* [405]:
(Summarized by) Dewick, P. M. (1984). *Natural Product Reports* 1, 459.
12. *Biosynthesis of prelunularic acid* [407]:
Abe, S., Kobayashi, M., Komura, H., and Ohta, Y. (1984). *Phytochemistry* 23, 1379.
13. *Indole alkaloid biosynthesis*:
(Summarized by) Herbert, R. B. (1985). *Natural Product Reports* 2, 168.
14. *Enzymology of indole alkaloid biosynthesis*:
(Summarized by) Hutchinson, C. R. (1986). *Natural Product Reports* 3, 136.
15. *Biosynthesis of camptothecin* [417]:
(Summarized by) Cai, J-C., and Hutchinson, C. R. (1983). In *The alkaloids*, (ed. A. Brossi), Vol. XXI, p. 118.
16. *Biosynthesis of quinine* [410]:
(Summarized by) Uskokovic, M. R. and Grethe, G. (1973). In *The alkaloids*, (ed. R. H. F. Manske), Vol. XIV, p. 209.
17. *Biosynthesis of ergot alkaloids*:
Shibuya, M., Chou, H-M., Fountoulakis, M., Hassam, S., Kim, S-U., Kobayashi, K., Otsuka, H., Rogalska, E., Cassady, J. M., and Floss, H. G. (1990). *J. Amer. Chem. Soc.* 112, 297.

Chapter 7

Most of the recommended reading has already been included in the first section, but it is worth re-emphasizing the value of two books:

Harborne, J. B. (1982). *Introduction to ecological biochemistry*, 2nd edn., Academic Press, London.
Haslam, E. (1985). *Metabolites and metabolism*, Oxford University Press, Oxford.

Several biosyntheses are mentioned in the chapter, and references for these are given below:

1. *Biosynthesis of coccinelline* [424]:
Braekman, J. C., Daloze, D., Hootele, C., Pasteels, J. M., and Tursch, B. (1975). *Tetrahedron* 31, 1541.
2. *Biosynthesis of cantharidine* [454]:
Hauff, S. A., Schmid, H., and Woggan, W-D. (1983). *Chem. Commun.*, 272.
3. *Biosynthesis of isoflavonoid phytoallexins*:
Smith, D. A. and Banks, S. W. (1986). *Phytochemistry* 25, 979.

Index

Taxol A   33069-62-4
Taxol B   71610-00-9